GLENCOE
MATHEMATICS

# Noteables™
## Interactive Study Notebook with FOLDABLES™

# Mathematics
## Applications and Concepts
### Course 2

**Contributing Author**
Dinah Zike

FOLDABLES™

**Consultant**
Douglas Fisher, PhD
Director of Professional Development
San Diego State University
San Diego, CA

Mc Graw Hill **Glencoe**

New York, New York   Columbus, Ohio   Chicago, Illinois   Peoria, Illinois   Woodland Hills, California

**Glencoe**

Send all inquiries to:
The McGraw-Hill Companies
8787 Orion Place
Columbus, OH  43240-4027

ISBN: 0-07-868215-0

*Mathematics: Applications and Concepts*, Course 2 *(Student Edition)*
*Noteables™: Interactive Study Notebook with Foldables™*

1 2 3 4 5 6 7 8 9 10 009 09 08 07 06 05 04

# Contents

# Contents

# Organizing Your Foldables

**FOLDABLES**™ Make this Foldable to help you organize and store your chapter Foldables. Begin with one sheet of 11″ × 17″ paper.

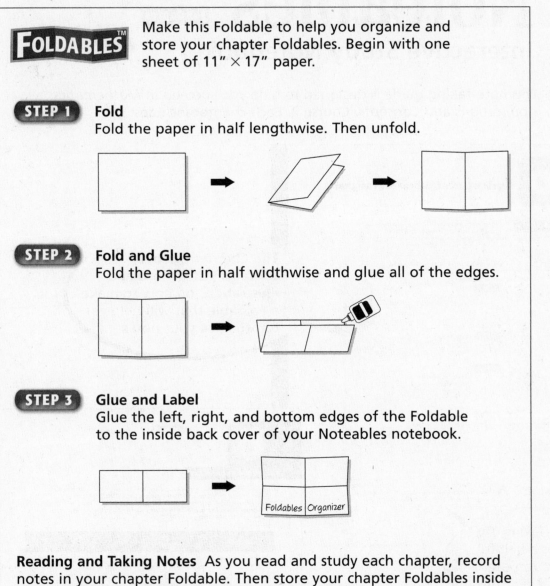

**STEP 1** **Fold**
Fold the paper in half lengthwise. Then unfold.

**STEP 2** **Fold and Glue**
Fold the paper in half widthwise and glue all of the edges.

**STEP 3** **Glue and Label**
Glue the left, right, and bottom edges of the Foldable to the inside back cover of your Noteables notebook.

Foldables | Organizer

**Reading and Taking Notes** As you read and study each chapter, record notes in your chapter Foldable. Then store your chapter Foldables inside this Foldable organizer.

# Using Your

# Noteables™
### with FOLDABLES

## Interactive Study Notebook

This note-taking guide is designed to help you succeed in *Mathematics: Applications and Concepts,* Course 2. Each chapter includes:

**CHAPTER 4**

**Algebra: Linear Equations and Functions**

**FOLDABLES** Use the instructions below to make a Foldable to help you organize your notes as you study the chapter. You will see Foldable reminders in the margin of this Interactive Study Notebook to help you in taking notes.

**Begin with a sheet of $8\frac{1}{2}$" × 11" paper.**

**STEP 1** **Fold**
Fold the short sides toward the middle.

**STEP 2** **Fold Again**
Fold the top to the bottom.

**STEP 3** **Cut**
Open. Cut along the second fold to make four tabs.

**STEP 4** **Label**
Label each of the tabs as shown.

**NOTE-TAKING TIP:** When you take notes, listen or read for main ideas. Then record those ideas in a simplified form for future reference.

*Mathematics: Applications and Concepts, Course 2* 8

The **Chapter Opener** contains instructions and illustrations on how to make a Foldable that will help you to organize your notes.

A **Note-Taking Tip** provides a helpful hint you can use when taking notes.

The **Build Your Vocabulary** table allows you to write definitions and examples of important vocabulary terms together in one convenient place.

**CHAPTER 4**

**BUILD YOUR VOCABULARY**

This is an alphabetical list of new vocabulary terms you will learn in Chapter 4. As you complete the study notes for the chapter, you will see Build Your Vocabulary reminders to complete each term's definition or description on these pages. Remember to add the textbook page number in the second column for reference when you study.

| Vocabulary Term | Found on Page | Definition | Description or Example |
|---|---|---|---|
| Addition Property of Equality | | | |
| Division Property of Equality | | | |
| domain | | | |
| function | | | |
| function table | | | |
| inequality | | | |
| inverse operations | | | |

Within each chapter, **Build Your Vocabulary** boxes will remind you to fill in this table.

84 *Mathematics: Applications and Concepts, Course 2*

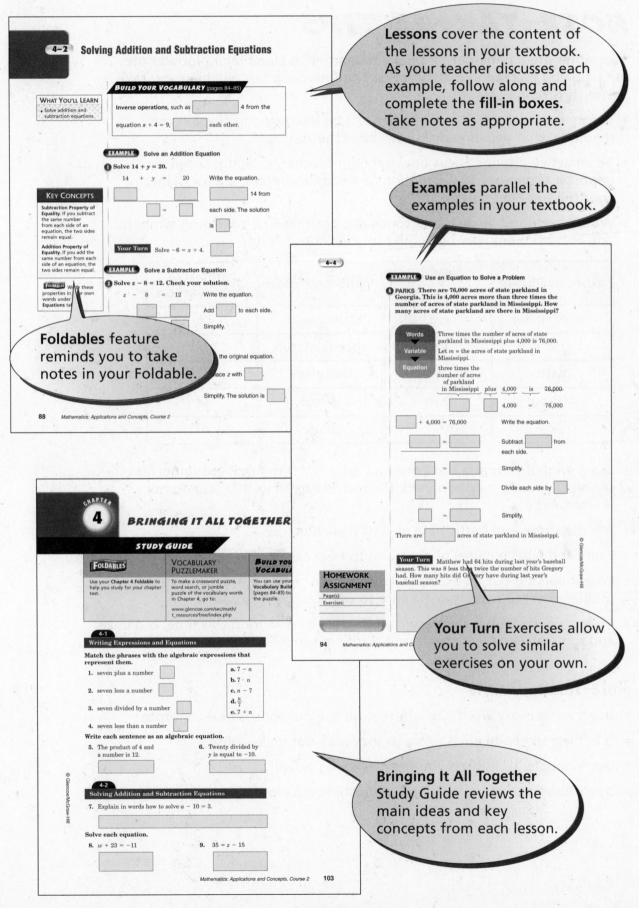

**4-2** Solving Addition and Subtraction Equations

**WHAT YOU'LL LEARN**
• Solve addition and subtraction equations.

**BUILD YOUR VOCABULARY** (pages 84–85)

Inverse operations, such as [ ] 4 from the

equation $x + 4 = 9$, [ ] each other.

**EXAMPLE** Solve an Addition Equation

❶ Solve $14 + y = 20$.

   $14 + y = 20$    Write the equation.

   [ ]        14 from
              each side. The solution
   [ ] = [ ]   is [ ]

**KEY CONCEPTS**

**Subtraction Property of Equality.** If you subtract the same number from each side of an equation, the two sides remain equal.

**Addition Property of Equality.** If you add the same number from each side of an equation, the two sides remain equal.

**FOLDABLES** Write these properties in your own words under Equations tab.

**Your Turn** Solve $-6 = x + 4$. [ ]

**EXAMPLE** Solve a Subtraction Equation

❷ Solve $z - 8 = 12$. Check your solution.

   $z - 8 = 12$    Write the equation.

              Add [ ] to each side.

              Simplify.

              the original equation.

              lace $z$ with [ ]

              Simplify. The solution is [ ]

88    Mathematics: Applications and Concepts, Course 2

---

**Lessons** cover the content of the lessons in your textbook. As your teacher discusses each example, follow along and complete the **fill-in boxes**. Take notes as appropriate.

**Examples** parallel the examples in your textbook.

**Foldables** feature reminds you to take notes in your Foldable.

---

**4-4**

**EXAMPLE** Use an Equation to Solve a Problem

❹ **PARKS** There are 76,000 acres of state parkland in Georgia. This is 4,000 acres more than three times the number of acres of state parkland in Mississippi. How many acres of state parkland are there in Mississippi?

**Words** → Three times the number of acres of state parkland in Mississippi plus 4,000 is 76,000.

**Variable** → Let $m$ = the acres of state parkland in Mississippi.

**Equation** →

three times the
number of acres
of parkland
in Mississippi    plus  4,000  is  76,000.

[ ]              4,000  =  76,000

[ ] + 4,000 = 76,000    Write the equation.

[ ] = [ ]    Subtract [ ] from each side.

[ ] = [ ]    Simplify.

[ ] = [ ]    Divide each side by [ ].

[ ] = [ ]    Simplify.

There are [ ] acres of state parkland in Mississippi.

**Your Turn** Matthew had 64 hits during last year's baseball season. This was 8 less than twice the number of hits Gregory had. How many hits did Gregory have during last year's baseball season?

**HOMEWORK ASSIGNMENT**

Page(s): _____
Exercises: _____

**Your Turn** Exercises allow you to solve similar exercises on your own.

94    Mathematics: Applications and C

---

**CHAPTER 4**

**BRINGING IT ALL TOGETHER**

**STUDY GUIDE**

| **FOLDABLES** | **VOCABULARY PUZZLEMAKER** | **BUILD YOU VOCABUL** |
|---|---|---|
| Use your Chapter 4 Foldable to help you study for your chapter test. | To make a crossword puzzle, word search, or jumble puzzle of the vocabulary words in Chapter 4, go to: www.glencoe.com/sec/math/t_resources/free/index.php | You can use your **Vocabulary Build** (pages 84–85) to the puzzle. |

**4-1**
**Writing Expressions and Equations**

Match the phrases with the algebraic expressions that represent them.

1. seven plus a number [ ]

2. seven less a number [ ]

3. seven divided by a number [ ]

4. seven less than a number [ ]

**a.** $7 - n$
**b.** $7 \cdot n$
**c.** $n - 7$
**d.** $\frac{n}{7}$
**e.** $7 + n$

Write each sentence as an algebraic equation.

5. The product of 4 and a number is 12.

6. Twenty divided by $y$ is equal to $-10$.

**4-2**
**Solving Addition and Subtraction Equations**

7. Explain in words how to solve $a - 10 = 3$.

Solve each equation.

8. $w + 23 = -11$

9. $35 = z - 15$

Mathematics: Applications and Concepts, Course 2    **103**

**Bringing It All Together** Study Guide reviews the main ideas and key concepts from each lesson.

---

*Mathematics: Applications and Concepts*, Course 2    **vii**

# NOTE-TAKING TIPS

Your notes are a reminder of what you learned in class. Taking good notes can help you succeed in mathematics. The following tips will help you take better classroom notes.

- Before class, ask what your teacher will be discussing in class. Review mentally what you already know about the concept.

- Be an active listener. Focus on what your teacher is saying. Listen for important concepts. Pay attention to words, examples, and/or diagrams your teacher emphasizes.

- Write your notes as clear and concise as possible. The following symbols and abbreviations may be helpful in your note-taking.

| Word or Phrase | Symbol or Abbreviation | Word or Phrase | Symbol or Abbreviation |
|---|---|---|---|
| for example | e.g. | not equal | $\neq$ |
| such as | i.e. | approximately | $\approx$ |
| with | w/ | therefore | $\therefore$ |
| without | w/o | versus | vs |
| and | + | angle | $\angle$ |

- Use a symbol such as a star ($\star$) or an asterisk (*) to emphasis important concepts. Place a question mark (?) next to anything that you do not understand.

- Ask questions and participate in class discussion.

- Draw and label pictures or diagrams to help clarify a concept.

- When working out an example, write what you are doing to solve the problem next to each step. Be sure to use your own words.

- Review your notes as soon as possible after class. During this time, organize and summarize new concepts and clarify misunderstandings.

## Note-Taking Don'ts

- **Don't** write every word. Concentrate on the main ideas and concepts.

- **Don't** use someone else's notes as they may not make sense.

- **Don't** doodle. It distracts you from listening actively.

- **Don't** lose focus or you will become lost in your note-taking.

# CHAPTER 1

# Decimal Patterns and Algebra

 Use the instructions below to make a Foldable to help you organize your notes as you study the chapter. You will see Foldable reminders in the margin of this Interactive Study Notebook to help you in taking notes.

**Begin with ten sheets of notebook paper.**

**STEP 1**  **Staple**
Staple the ten sheets together to form a booklet.

**STEP 2**  **Cut Tabs**
On the second page, make the top tab the width of the white space. On the third page, make the tab 2 lines longer, and so on.

**STEP 3**  **Label**
Write the chapter title on the cover and label each tab with the lesson number.

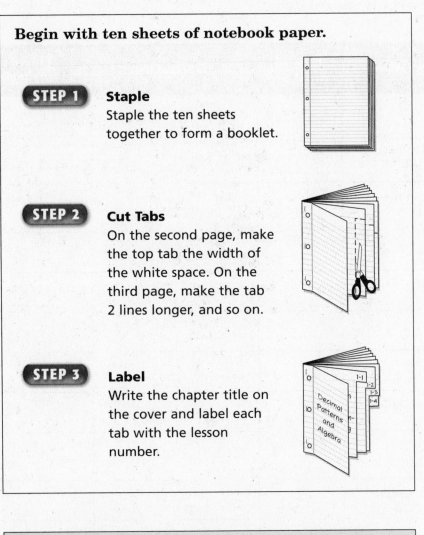

**NOTE-TAKING TIP:** When taking notes, it is often a good idea to write a summary of the lesson in your own words. Be sure to paraphrase key points.

## BUILD YOUR VOCABULARY

This is an alphabetical list of new vocabulary terms you will learn in Chapter 1. As you complete the study notes for the chapter, you will see Build Your Vocabulary reminders to complete each term's definition or description on these pages. Remember to add the textbook page number in the second column for reference when you study.

| Vocabulary Term | Found on Page | Definition | Description or Example |
|---|---|---|---|
| algebra | | | |
| algebraic expression [al-juh-BRAY-ihk] | | | |
| arithmetic sequence [air-ith-MEH-tik] | | | |
| base | | | |
| coefficient | | | |
| constant | | | |
| cubed | | | |
| defining the variable | | | |
| equation [ih-KWAY-zhuhn] | | | |
| equivalent expressions | | | |
| evaluate | | | |
| exponent | | | |
| exponential form | | | |

| Vocabulary Term | Found on Page | Definition | Description or Example |
|---|---|---|---|
| factors | | | |
| geometric sequence | | | |
| gram | | | |
| kilogram | | | |
| liter [LEE-tuhr] | | | |
| meter | | | |
| metric system | | | |
| numerical expression | | | |
| order of operations | | | |
| powers | | | |
| properties | | | |
| scientific notation | | | |
| sequence | | | |
| solution | | | |
| solving an equation | | | |
| squared | | | |
| standard form | | | |
| term | | | |
| variable | | | |

**1–1** **A Plan for Problem Solving**

**EXAMPLE** Use the Four-Step Plan

---

**WHAT YOU'LL LEARN**

• Solve problems using the four-step plan.

---

**1** SPENDING A can of soda holds 12 fluid ounces. A 2-liter bottle holds about 67 fluid ounces. If a pack of six cans costs the same as a 2-liter bottle, which is the better buy?

**EXPLORE** You know the number of fluid ounces of soda in one can of soda. You need to know the number of fluid ounces of soda in a pack of six cans.

**PLAN** You can find the number of fluid ounces of soda in a pack of six cans by [_____] the number of fluid ounces in one can by [_____].

**SOLVE** $12 \times$ [_____] $=$ [_____]

There are [_____] fluid ounces of soda in a pack of six cans. The number of fluid ounces of soda in a 2-liter bottle is about [_____]. Therefore, the

[_____] is the better buy because you get more soda for the same price.

**EXAMINE** The answer makes sense based on the facts given in the problem.

---

**FOLDABLES**

**ORGANIZE IT**

Summarize the four-step problem-solving plan on the Lesson 1-1 page of your Foldable.

Decimal Patterns and Algebra
1-1 1-2 1-3 1-4

---

**Your Turn** The sixth grade class at Meadow Middle School is taking a field trip to the local zoo. There will be 142 students plus 12 adults going on the trip. If each school bus can hold 48 people, how many buses will be needed for the field trip?

[_____]

---

**EXAMPLE** Use a Strategy in the Four-Step Plan

**②** POPULATION For every 100,000 people in the United States, there are 5,750 radios. For every 100,000 people in Canada, there are 323 radios. Suppose Sheamus lives in Des Moines, Iowa, and Alex lives in Windsor, Ontario. Both cities have about 200,000 residents. About how many more radios are there in Sheamus's city than in Alex's city?

**EXPLORE** You know the approximate number of radios per 100,000 people in both Sheamus's city and Alex's city.

**PLAN** You can find the approximate number of radios in each city by ☐ the estimate per 100,000 people by two to get an estimate per 200,000 people. Then, ☐ to find how many more radios there are in Des Moines than in Windsor.

**SOLVE** Des Moines: $5{,}750 \times 2 = $ ☐

Windsor: $323 \times 2 = $ ☐

☐ $-$ ☐ $=$ ☐

So, Des Moines has about ☐ more radios than Windsor.

**EXAMINE** Based on the information given in the problem, the answer seems to be reasonable.

---

**Your Turn** Ben borrows a 500-page book from the library. On the first day, he reads 24 pages. On the second day, he reads 39 pages and on the third day he reads 54 pages. If Ben follows the same pattern of number of pages read for seven days, will he have finished the book at the end of the week?

---

## KEY CONCEPTS

**Problem-Solving Strategies**

- guess and check
- look for a pattern
- make an organized list
- draw a diagram
- act it out
- solve a simpler problem
- use a graph
- work backward
- eliminate possibilities
- estimate reasonable answers
- use logical reasoning
- make a model

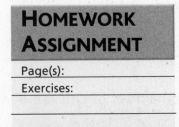

## HOMEWORK ASSIGNMENT

Page(s): _____
Exercises: _____

# Powers and Exponents

## BUILD YOUR VOCABULARY (pages 2–3)

Two or more numbers that are multiplied together to form a [____] are called **factors**.

The **exponent** tells how many times the base is used as a [____].

The **base** is the common [____].

Numbers expressed using [____] are called **powers**.

Five to the [____] power is five **squared**.

Four to the [____] power is four **cubed**.

**FOLDABLES**

## ORGANIZE IT
On the Lesson 1-2 page of your Foldable, explain the difference between the terms power and exponent.

*Decimal Patterns and Algebra*

**EXAMPLES** Write Powers as Products

**Write each power as a product of the same factor.**

**1** $8^4$

The base is [____]. The exponent [____] means that 8 is used as a factor [____] times. $8^4 =$ [____]

**2** $4^6$

The base is [____]. The exponent [____] means that [____] is used as a factor six times. $4^6 =$ [____]

**Your Turn** Write each power as a product of the same factor.

a. $3^6$ [____]     b. $7^3$ [____]

**BUILD YOUR VOCABULARY** (pages 2–3)

You can **evaluate**, or find the [ ] of, [ ]

by multiplying the factors.

Numbers written [ ] are in

**standard form**.

Numbers written [ ] are in

**exponential form**.

## WRITE IT

Explain how you would use a calculator to evaluate a power.

_____
_____
_____
_____
_____

**EXAMPLES** Write Powers in Standard Form

**Evaluate each expression.**

❸ $8^3$ = [ ] = [ ]

❹ $6^4$ = [ ] = [ ]

**Your Turn** Evaluate each expression.

**a.** $4^4$ = [ ]　　**b.** $5^5$ = [ ]

**EXAMPLE** Write Numbers in Exponential Form

❺ Write $9 \cdot 9 \cdot 9 \cdot 9 \cdot 9 \cdot 9$ in exponential form.

9 is the [ ]. It is used as a factor [ ] times.

So the exponent is [ ].

[ ] = [ ]

**Your Turn** Write $3 \cdot 3 \cdot 3 \cdot 3 \cdot 3$ in exponential form.

[ ]

## HOMEWORK ASSIGNMENT

Page(s): _____
Exercises: _____

# Order of Operations

• Evaluate expressions using the order of operations.

**BUILD YOUR VOCABULARY** (pages 2–3)

The expressions $4 \cdot 6 - (5 + 7)$ and $8 \cdot (9 - 3) + 4$ are

[               ] expressions.

**Order of operations** are [           ] that

ensure that numerical expressions have only one value.

**EXAMPLES** Evaluate Expressions

## KEY CONCEPT

**Order of Operations**

1. Do all operations within grouping symbols first.

2. Evaluate all powers before other operations.

3. Multiply and divide in order from left to right.

4. Add and subtract in order from left to right.

**FOLDABLES** Be sure to include the order of operations on the Lesson 1-3 page of your Foldable.

Evaluate each expression.

**1** $27 - (18 + 2)$

$27 - (18 + 2) = 27 - $ [     ]          Add 18 and 2.

$= $ [     ]          Subtract 20 from 27.

**2** $15 + 5 \cdot 3 - 2$

$15 + 5 \cdot 3 - 2 = 15 + $ [     ] $- 2$          Multiply 5 and 3.

$= $ [     ] $- 2$          Add 15 and 15.

$= $ [     ]          Subtract 2 from 30.

**Your Turn** Evaluate each expression.

**a.** $45 - (26 + 3)$

**b.** $32 - 3 \cdot 7 + 4$

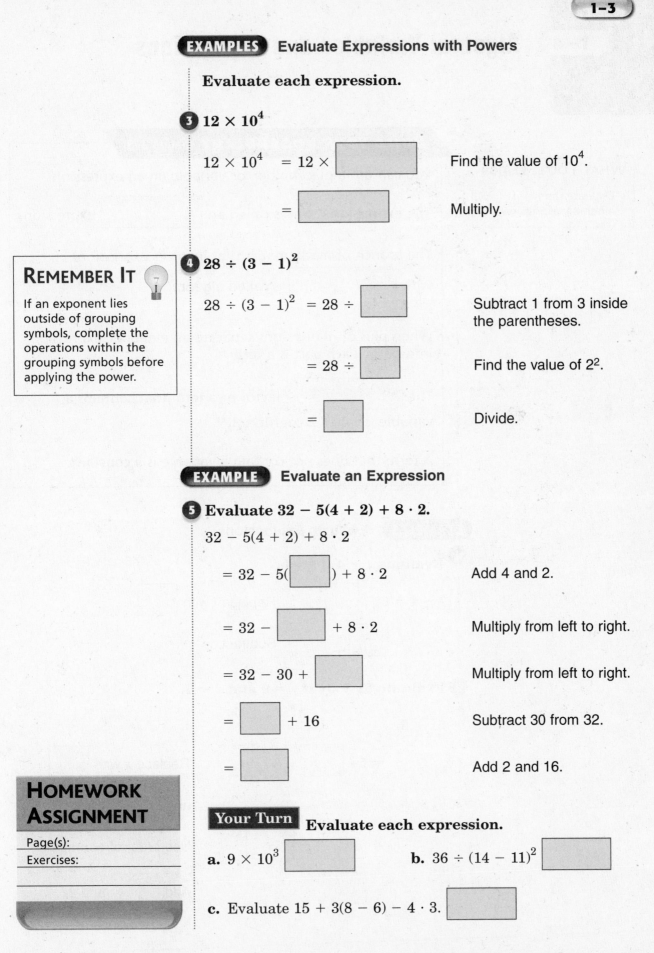

**EXAMPLES** Evaluate Expressions with Powers

Evaluate each expression.

**3** $12 \times 10^4$

$12 \times 10^4 \quad = 12 \times \boxed{\phantom{xxxx}}$   Find the value of $10^4$.

$= \boxed{\phantom{xxxx}}$   Multiply.

**REMEMBER IT**

If an exponent lies outside of grouping symbols, complete the operations within the grouping symbols before applying the power.

**4** $28 \div (3 - 1)^2$

$28 \div (3 - 1)^2 = 28 \div \boxed{\phantom{xx}}$   Subtract 1 from 3 inside the parentheses.

$= 28 \div \boxed{\phantom{xx}}$   Find the value of $2^2$.

$= \boxed{\phantom{xx}}$   Divide.

**EXAMPLE** Evaluate an Expression

**5** Evaluate $32 - 5(4 + 2) + 8 \cdot 2$.

$32 - 5(4 + 2) + 8 \cdot 2$

$= 32 - 5(\boxed{\phantom{xx}}) + 8 \cdot 2$   Add 4 and 2.

$= 32 - \boxed{\phantom{xx}} + 8 \cdot 2$   Multiply from left to right.

$= 32 - 30 + \boxed{\phantom{xx}}$   Multiply from left to right.

$= \boxed{\phantom{xx}} + 16$   Subtract 30 from 32.

$= \boxed{\phantom{xx}}$   Add 2 and 16.

**HOMEWORK ASSIGNMENT**

Page(s):

Exercises:

**Your Turn** Evaluate each expression.

**a.** $9 \times 10^3$ $\boxed{\phantom{xxxx}}$   **b.** $36 \div (14 - 11)^2$ $\boxed{\phantom{xxxx}}$

**c.** Evaluate $15 + 3(8 - 6) - 4 \cdot 3$. $\boxed{\phantom{xxxx}}$

# Algebra: Variables and Expressions

**BUILD YOUR VOCABULARY** (pages 2–3)

**WHAT YOU'LL LEARN**

- Evaluate simple algebraic expressions.

You can use a placeholder, or **variable,** in an expression.

The expression $7 + n$ is called an [              ] expression.

The branch of mathematics that involves expressions

with [              ] is called **algebra**.

When plus or minus signs separate an algebraic expression into parts, each part is a **term**.

The [              ] factor of a term that contains a variable is called a **coefficient**.

A term that does not contain a variable is a **constant**.

**EXAMPLES**   Evaluate Expressions

**1** Evaluate $t - 4$ if $t = 6$.

$t - 4 = 6 - $ [     ]     Replace $t$ with [     ].

$\quad = $ [     ]     Subtract.

**2** Evaluate $5x + 3y$ if $y = 9$ and $x = 7$.

$5x + 3y$

$= 5 \cdot$ [     ] $+ 3 \cdot$ [     ]     Replace $x$ with [     ]

and [     ] with 9.

$= $ [     ] $+$ [     ]     Use order of operations.

$= $ [     ]     Add [     ] and 27.

**3** **Evaluate $\frac{rs}{4}$ if $r = 7$ and $s = 12$.**

$$\frac{rs}{4} = \frac{7 \cdot \boxed{\phantom{xx}}}{4}$$

Replace $r$ with $\boxed{\phantom{xx}}$ and $s$ with $\boxed{\phantom{xx}}$.

$$= \frac{\boxed{\phantom{xx}}}{4}$$

Multiply 7 and $\boxed{\phantom{xx}}$.

$$= \boxed{\phantom{xx}}$$

$\boxed{\phantom{xxxxxxxx}}$

**4** **Evaluate $5 + a^2$ if $a = 5$.**

$$5 + a^2 = 5 + 5^2$$

Replace $a$ with $\boxed{\phantom{xx}}$.

$$= 5 + \boxed{\phantom{xx}}$$

Use $\boxed{\phantom{xxxxxxxx}}$.

$$= \boxed{\phantom{xx}}$$

Add $\boxed{\phantom{xx}}$ and 25.

**Your Turn** **Evaluate each expression.**

**a.** $7 + m$ if $m = 4$.

**b.** $4a - 2b$ if $a = 9$ and $b = 6$.

**c.** $\frac{gh}{6}$ if $g = 8$ and $h = 9$.

**d.** $24 - s^2$ if $s = 3$.

**FOLDABLES™**

## ORGANIZE IT

Record and evaluate an example of a simple algebraic expression on the Lesson 1-4 page of your Foldable.

Decimal Patterns and Algebra

1-1
1-2
1-3
1-4

## HOMEWORK ASSIGNMENT

Page(s): _____

Exercises: _____

_____

_____

## 1–5  Algebra: Equations

**WHAT YOU'LL LEARN**

• Solve equations using mental math.

**BUILD YOUR VOCABULARY** (pages 2–3)

An **equation** is a [ ] in mathematics that contains an equal sign.

The **solution** of an equation is a number that makes the sentence [ ].

The process of finding a [ ] is called **solving an equation.**

When you choose a [ ] to represent one of the unknowns in an equation, you are **defining the variable.**

---

**FOLDABLES**

**ORGANIZE IT**

On the Lesson 1-5 page of your Foldable, record and solve an example of an algebraic expression.

Decimal Patterns and Algebra

1-1
1-2
1-3
1-4

---

**EXAMPLE**  Solve an Equation Mentally

❶ Solve $\frac{x}{2} = 3$ mentally.

$\frac{x}{2} = 3$  Write the equation.

$\dfrac{\boxed{\phantom{x}}}{2} = 3$  You know that $\frac{6}{2}$ is $\boxed{\phantom{x}}$.

$\boxed{\phantom{x}} = 3$  Simplify.

The solution is $\boxed{\phantom{x}}$.

**Your Turn**  Solve $p - 6 = 11$ mentally.

---

**EXAMPLE** Graph the Solution of an Equation

**②** **Graph the solution of $\frac{x}{2} = 3$.**

The solution of $\frac{x}{2} = 3$ is 6. Locate the point named by the

solution on a [ ]. Then draw a dot at the

solution, 6.

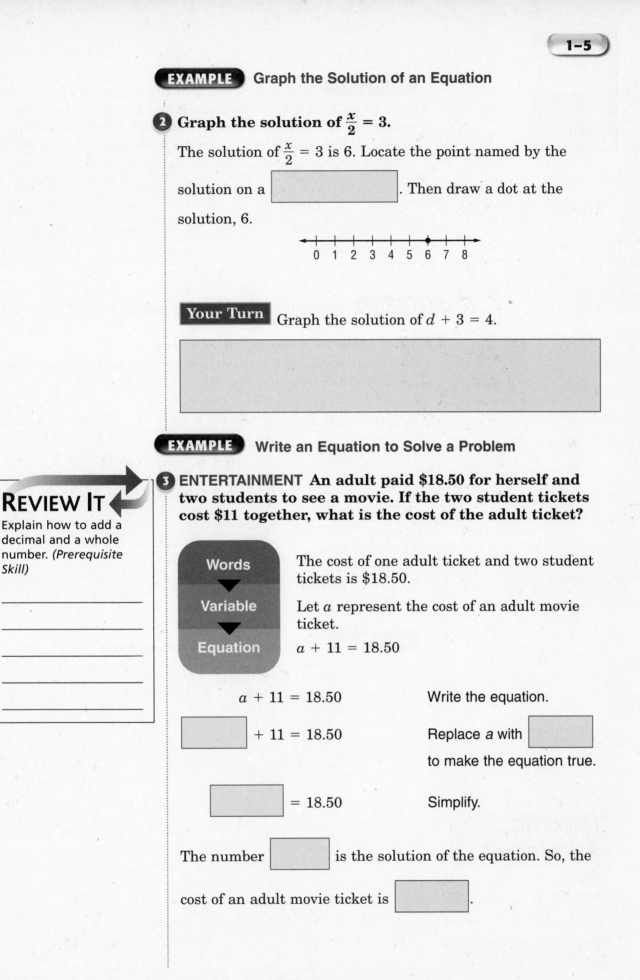

**Your Turn** Graph the solution of $d + 3 = 4$.

**EXAMPLE** Write an Equation to Solve a Problem

**③** **ENTERTAINMENT An adult paid $18.50 for herself and two students to see a movie. If the two student tickets cost $11 together, what is the cost of the adult ticket?**

| | |
|---|---|
| **Words** ▼ | The cost of one adult ticket and two student tickets is $18.50. |
| **Variable** ▼ | Let $a$ represent the cost of an adult movie ticket. |
| **Equation** | $a + 11 = 18.50$ |

$a + 11 = 18.50$      Write the equation.

[ ] $+ 11 = 18.50$      Replace $a$ with [ ]
to make the equation true.

[ ] $= 18.50$      Simplify.

The number [ ] is the solution of the equation. So, the

cost of an adult movie ticket is [ ].

**REVIEW IT**

Explain how to add a decimal and a whole number. *(Prerequisite Skill)*

_____

_____

_____

_____

_____

**Your Turn** Julie spends $9.50 at the ice cream parlor. She buys a hot fudge sundae for herself and ice cream cones for each of the three friends who are with her. Find the cost of Julie's sundae if the three ice cream cones together cost $6.30.

**EXAMPLE** Find a Solution of an Equation

**4** **What value of $x$ is a solution of $12 - x = 5$?**

Substitute a value for $x$ to determine which value makes the left side of the equation equivalent to the right side.

Replace $x$ with 5.       Replace $x$ with 6.       Replace $x$ with 7.

$12 - x = 5$              $12 - x = 5$              $12 - x = 5$

$12 - 5 \stackrel{?}{=} 5$       $12 - 6 \stackrel{?}{=} 5$       $12 - 7 \stackrel{?}{=} 5$

☐ $\neq 5$ false       ☐ $\neq 5$ false       ☐ $= 5$ true ✔

The value ☐ makes the equation true.

So, ☐ is the solution of $12 - x = 5$.

**Your Turn** What value of $x$ is a solution of $x - 13 = 22$?

# Algebra: Properties

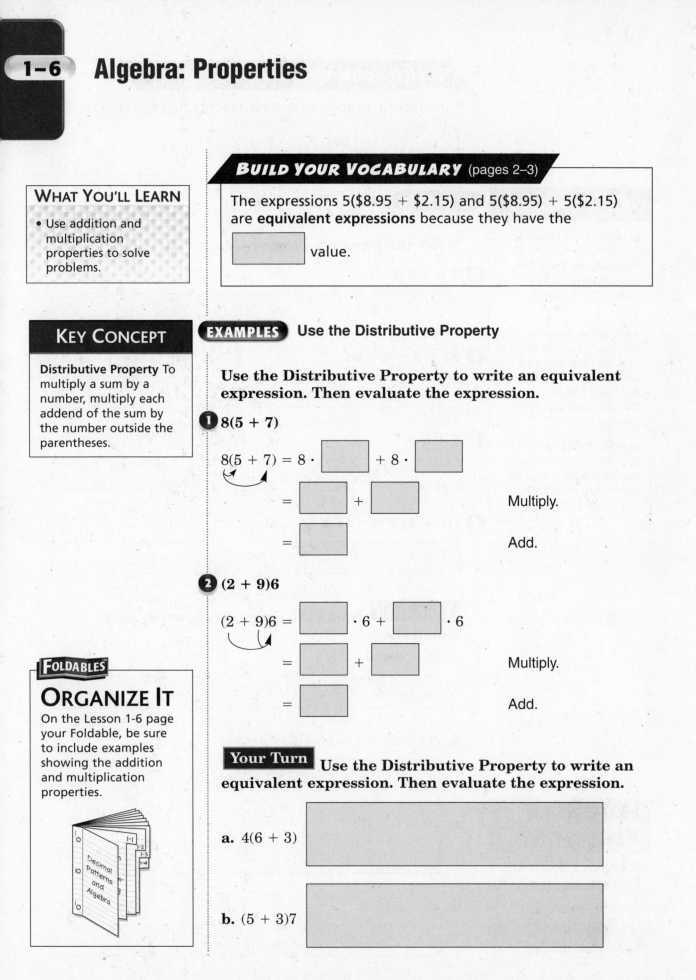

- Use addition and multiplication properties to solve problems.

**BUILD YOUR VOCABULARY** (pages 2–3)

The expressions 5($8.95 + $2.15) and 5($8.95) + 5($2.15) are **equivalent expressions** because they have the

[          ] value.

## KEY CONCEPT

**Distributive Property** To multiply a sum by a number, multiply each addend of the sum by the number outside the parentheses.

**EXAMPLES** Use the Distributive Property

**Use the Distributive Property to write an equivalent expression. Then evaluate the expression.**

**❶** $8(5 + 7)$

$8(5 + 7) = 8 \cdot \boxed{\phantom{xx}} + 8 \cdot \boxed{\phantom{xx}}$

$\qquad = \boxed{\phantom{xx}} + \boxed{\phantom{xx}}$   Multiply.

$\qquad = \boxed{\phantom{xx}}$   Add.

**❷** $(2 + 9)6$

$(2 + 9)6 = \boxed{\phantom{xx}} \cdot 6 + \boxed{\phantom{xx}} \cdot 6$

$\qquad = \boxed{\phantom{xx}} + \boxed{\phantom{xx}}$   Multiply.

$\qquad = \boxed{\phantom{xx}}$   Add.

**FOLDABLES**

## ORGANIZE IT

On the Lesson 1-6 page your Foldable, be sure to include examples showing the addition and multiplication properties.

Decimal Patterns and Algebra

**Your Turn** Use the Distributive Property to write an equivalent expression. Then evaluate the expression.

**a.** $4(6 + 3)$

**b.** $(5 + 3)7$

In algebra, **properties** are statements that are true for any

[          ] or [          ] .

## KEY CONCEPTS

**Commutative Property**
The order in which two numbers are added or multiplied does not change their sum or product.

**Associative Property**
The way in which three numbers are grouped when they are added or multiplied does not change their sum or product.

**Identity Property** The sum of an addend and zero is the addend. The product of a factor and one is the factor.

**EXAMPLES** Identify Properties

**Name the property shown by each statement.**

**3** $7 = 1 \times 7$

**4** $24 + 5 = 5 + 24$

**5** $7 + 0 = 7$

**6** $(11 \times 4) \times 8 = 11 \times (4 \times 8)$

**Your Turn** Name the property shown by each statement.

**a.** $15 + 9 = 9 + 15$

**b.** $(14 + 5) + 3 = 14 + (5 + 3)$

**c.** $10 = 1 \times 10$

**d.** $20 = 20 + 0$

## HOMEWORK ASSIGNMENT

Page(s):

Exercises:

# Sequences

## WHAT YOU'LL LEARN

- Recognize and extend patterns for sequences.

---

**BUILD YOUR VOCABULARY** (pages 2–3)

A **sequence** is an [＿＿＿＿] list of [＿＿＿＿].

Each number in a [＿＿＿＿] is called a **term**.

In an **arithmetic sequence**, each term is found by [＿＿＿＿] the same number to the [＿＿＿＿] term.

In a **geometric sequence**, each term is found by [＿＿＿＿] the same [＿＿＿＿] by the previous term.

---

## FOLDABLES

## ORGANIZE IT

Write an example of an arithmetic and a geometric sequence on the Lesson 1-7 page of your Foldable.

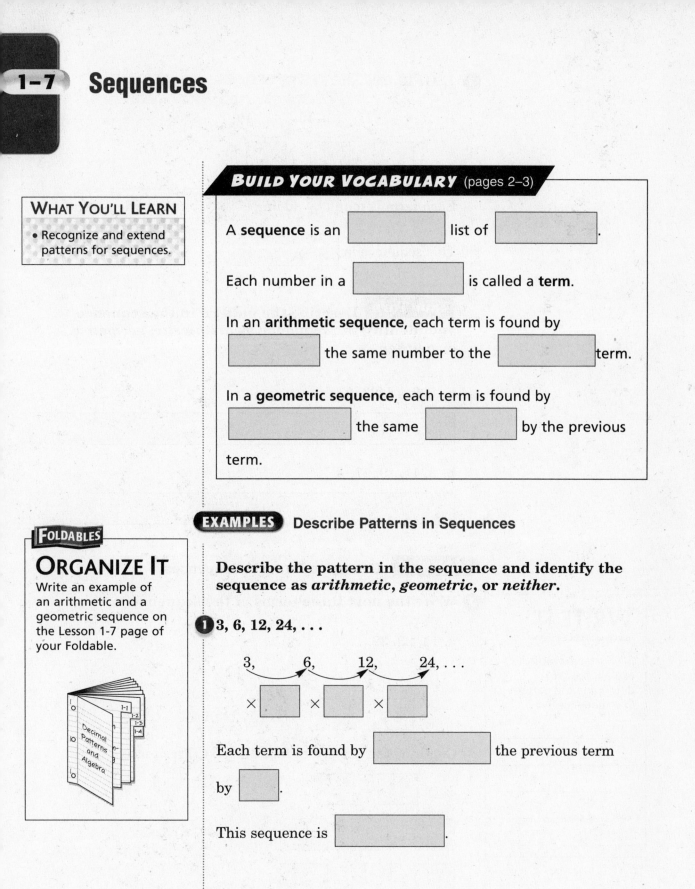

---

**EXAMPLES** Describe Patterns in Sequences

Describe the pattern in the sequence and identify the sequence as *arithmetic*, *geometric*, or *neither*.

**①** 3, 6, 12, 24, . . .

3,　　6,　　12,　　24, . . .

× [＿]　× [＿]　× [＿]

Each term is found by [＿＿＿＿] the previous term

by [＿].

This sequence is [＿＿＿＿].

**2** 7, 11, 15, 19, . . .

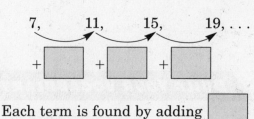

Each term is found by adding [ ] to the previous term.

This sequence is [ ].

**Your Turn** Describe the pattern in the sequence and identify the sequence as *arithmetic*, *geometric*, or *neither*.

**a.** 5, 9, 18, 22, 31, . . .

**b.** 3, 11, 19, 27, . . .

**EXAMPLES** Determine Terms in Sequences

**3** Write the next three terms of the sequence.

5, 14, 23, 32, . . .

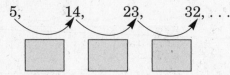

Continue the pattern to find the next three terms.

32 + [ ] = [ ]

[ ] + [ ] = [ ]

[ ] + [ ] = [ ]

The next three terms are [ ].

**WRITE IT**

In your own words, explain how to determine the pattern in a sequence.

_____

_____

_____

_____

_____

**4** 0.2, 1.2, 7.2, 43.2, . . .

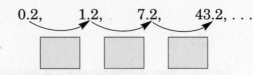

0.2,     1.2,     7.2,     43.2, . . .

Continue the pattern to find the next three terms.

43.2 · ☐ = ☐

☐ · ☐ = ☐

☐ · ☐ = ☐

The next three terms are ☐ .

**Your Turn** Write the next three terms of the sequence.

**a.** 12, 17, 22, 27, . . .

**b.** 3, 12, 48, 192, . . .

---

## REVIEW IT

Explain how you know where to place the decimal point when multiplying by a decimal.

_____

_____

_____

_____

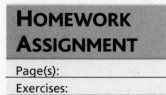

## HOMEWORK ASSIGNMENT

Page(s):

Exercises:

_____

_____

# Measurement: The Metric System

## BUILD YOUR VOCABULARY (pages 2–3)

The [          ] is the base unit of length in the

**metric system**.

A **meter** is about the distance from the floor to a

doorknob or a little more than a [          ].

**FOLDABLES**

## ORGANIZE IT

On the Lesson 1-8 page of the Foldable, write the meanings of the metric prefixes *kilo–*, *centi–*, and *milli–*.

Decimal Patterns and Algebra

1-1
1-2
1-3
1-4

**EXAMPLES**  Convert Units of Length

**1** Complete 28 cm = _?_ mm.

To convert from centimeters to millimeters, [          ]

by [          ].

28 × [          ] = [          ]

So, 28 cm = [          ] mm.

**2** Complete 438 cm = _?_ mm.

To convert from centimeters to meters, [          ] by [          ].

438 ÷ [          ] = [          ]

So, 438 cm = [          ] m.

**Your Turn** Complete.

**a.** 3,400 mm = _?_ cm

**b.** 7.5 m = _?_ cm

**BUILD YOUR VOCABULARY** (pages 2–3)

A paperclip has a [ ] of about one **gram**.

The base unit of mass in the metric system is the **kilogram**, which is equivalent to [ ] grams.

**EXAMPLES**  Convert Units of Mass

**3** Complete 72 g = $\frac{?}{}$ mg.

To convert from grams to milligrams, [ ]

by [ ].

$72 \times$ [ ] = [ ]

So, 72 g = [ ] mg.

## WRITE IT

Explain how you can multiply a number by a power of ten.

_____

_____

_____

_____

**4** Complete 202 g = $\frac{?}{}$ kg.

To convert from grams to kilograms, [ ]

by [ ].

$202 \div$ [ ] = [ ]

So, 202 g = [ ].

**Your Turn** Complete.

**a.** 4,550 mg = $\frac{?}{}$ g

**b.** 6.25 kg = $\frac{?}{}$ g

**BUILD YOUR VOCABULARY** (pages 2–3)

The **liter** is widely used and accepted as the [ ]

measure for [ ].

**EXAMPLES** Convert Units of Capacity

**5** Complete 2 L = ___?___ mL.

To convert from liters to milliliters, [ ]

by [ ].

2 × [ ] = [ ]

So, 2 L = [ ].

**6** Complete 2.4 kL = ___?___ L.

To convert from kiloliters to liters, [ ]

by [ ].

2.4 × [ ] = [ ]

So, 2.4 kL = [ ].

**Your Turn** Complete.

**a.** 450 mL = ___?___ L

[ ]

**b.** 95.3 L = ___?___ kL

[ ]

**HOMEWORK ASSIGNMENT**

Page(s): _____

Exercises: _____

_____

_____

# 1–9   Scientific Notation

## WHAT YOU'LL LEARN

- Write numbers greater than 100 in scientific notation and in standard form.

## KEY CONCEPT

**Scientific Notation** A number is expressed in scientific notation when it is written as the product of a number and a power of ten. The number must be greater than or equal to 1 and less than 10.

### BUILD YOUR VOCABULARY (pages 2–3)

Numbers like 4.2 million can be written in **scientific**

**notation** by using a power of [ ].

### EXAMPLES   Write a Number in Standard Form

**1** Write $3.4 \times 10^6$ in standard form.

$3.4 \times 10^6 = 3.4 \times$ [ ]          $10^6 =$ [ ]

$= 3400000$          Move the decimal point

[ ] places to the

[ ].

$=$ [ ]

**Your Turn**  Write $7.3 \times 10^4$ in standard form.

[ ]

### EXAMPLES   Write a Number in Scientific Notation

**2** Write 428,000 in scientific notation.

$428,000 =$ [ ] $\times 100,000$   Move the decimal point

[ ] places to find a number

between 1 and [ ].

$=$ [ ] $\times$ [ ]

$=$ [ ]

*Mathematics: Applications and Concepts, Course 2*      **23**

**Your Turn** Write 1,750,000 in scientific notation.

<br>

**EXAMPLE** Compute with Large Numbers

**3** ASTRONOMY Pluto's maximum distance from Earth is about 4.6 billion miles. One mile is equal to 5,280 feet. Find the approximate maximum distance of Pluto from Earth in feet.

To find the approximate maximum distance in feet,

[          ] 4.6 billion by [          ].

4.6 billion = [          ]

Using a calculator, you find that

[          ] × [          ] is $2.4288 \times 10^{13}$.

The approximate maximum distance of Pluto from Earth is about [          ].

**Your Turn** In order to build a new shopping center, 553,000 bricks were used. A new shopping mall built across town used twelve times as many bricks. About how many bricks were used to build the new shopping mall?

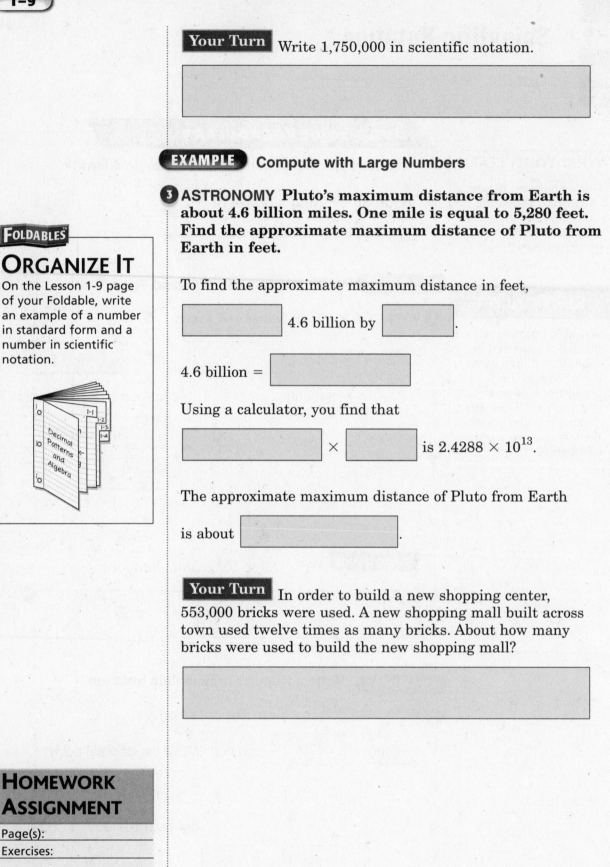

**HOMEWORK ASSIGNMENT**

Page(s):

Exercises:

# CHAPTER 1

# BRINGING IT ALL TOGETHER

## STUDY GUIDE

| **FOLDABLES™** | VOCABULARY PUZZLEMAKER | **BUILD YOUR VOCABULARY** |
|---|---|---|
| Use your **Chapter 1 Foldable** to help you study for your chapter test. | To make a crossword puzzle, word search, or jumble puzzle of the vocabulary words in Chapter 1, go to: www.glencoe.com/sec/math/t_resources/free/index.php | You can use your completed **Vocabulary Builder** (*pages 2–3*) to help you solve the puzzle. |

### 1-1
### A Plan for Problem Solving

**Underline the correct term to complete each sentence.**

1. The (*Plan, Solve*) step is the step of the four-step plan in which you decide which strategy you will use to solve the problem.

2. According to the four-step plan, if your answer is not correct, you should (*estimate the answer, make a new plan and start again*).

3. Once you solve a problem, make sure your solution contains any appropriate (*strategies, units or labels*).

### 1-2
### Powers and Exponents

**Identify the exponent in each expression.**

4. $5^8$ ☐

5. $8^3$ ☐

**Evaluate each expression.**

6. $4^3$ ☐

7. $8^5$ ☐

**Complete the sentence.**

8. Numbers written with exponents are in ☐ form, whereas numbers written without exponents are in ☐ form.

**1-3**

## Order of Operations

Evaluate each expression.

**9.** $9 + 18 \div 6$

**10.** $(7-4)^2 \div 3$

**11.** $2 \times 4^2 \div 4 - 1$

**12.** $8 + 2(9 - 5) - (2 \cdot 3)$

**1-4**

## Algebra: Equations

Evaluate each expression if $a = 5$ and $b = 6$.

**13.** $b - 2$

**14.** $2a + 3b$

**15.** $\dfrac{ab}{5}$

**16.** $a^2 - 3b$

**1-5**

## Algebra: Variables and Expressions

Solve each equation mentally.

**17.** $5 + b = 12$

**18.** $h - 6 = 3$

**19.** $12 \cdot 4 = n$

**20.** $2 = \dfrac{x}{4}$

**21.** $9t = 54$

**22.** $35 \div c = 7$

## 1-6
## Algebra: Properties

**Match the statement with the property it shows.**

**23.** $5 + (3 + 6) = (5 + 3) + 6$ ⬜    **a.** Distributive Property

**24.** $8 + 0 = 8$ ⬜    **b.** Commutative Property of Addition

**25.** $4(7 - 2) = 4(7) - 4(2)$ ⬜    **c.** Associative Property of Addition

**26.** $10 + 9 = 9 + 10$ ⬜    **d.** Identity Property of Addition

## 1-7
## Sequences

**Complete the sentence.**

**27.** In an arithmetic sequence, each term is found by ⬜ the same number to the previous term.

**28.** In a geometric sequence, each term is found by

⬜ the same number by the previous term.

**What is the next term in each of the following sequences?**

**29.** $1, 5, 25, \ldots$ ⬜       **30.** $7, 10, 13, \ldots$ ⬜

## 1-8
## Measurement: The Metric System

**Complete.**

**31.** $4.3 \text{ cm} = $ ⬜ mm       **32.** $42.7 \text{ g} = $ ⬜ mg

**33.** $690 \text{ mL} = $ ⬜ L

**34.** Write $3.5 \times 10^6$ in standard form. ⬜

**35.** Write 89,400,000 in scientific notation. ⬜

# ARE YOU READY FOR THE CHAPTER TEST?

Visit **msmath2.net** to access your textbook, more examples, self-check quizzes, and practice tests to help you study the concepts in Chapter 1.

**Check the one that applies. Suggestions to help you study are given with each item.**

☐ **I completed the review of all or most lessons without using my notes or asking for help.**

- You are probably ready for the Chapter Test.

- You may want to take the Chapter 1 Practice Test on page 49 of your textbook as a final check.

☐ **I used my Foldables or Study Notebook to complete the review of all or most lessons.**

- You should complete the Chapter 1 Study Guide and Review on pages 46–48 of your textbook.

- If you are unsure of any concepts or skills, refer back to the specific lesson(s).

- You may want to take the Chapter 1 Practice Test on page 49 of your textbook.

☐ **I asked for help from someone else to complete the review of all or most lessons.**

- You should review the examples and concepts in your Study Notebook and Chapter 1 Foldables.

- Then complete the Chapter 1 Study Guide and Review on pages 46–48 of your textbook.

- If you are unsure of any concepts or skills, refer back to the specific lesson(s).

- You may also want to take the Chapter 1 Practice Test on page 49 of your textbook.

Student Signature

Parent/Guardian Signature

Teacher Signature

# Statistics: Analyzing Data

Use the instructions below to make a Foldable to help you organize your notes as you study the chapter. You will see Foldable reminders in the margin of this Interactive Study Notebook to help you in taking notes.

**Statistics** Make this Foldable to help you organize information about analyzing data. Begin with eight sheets of notebook paper.

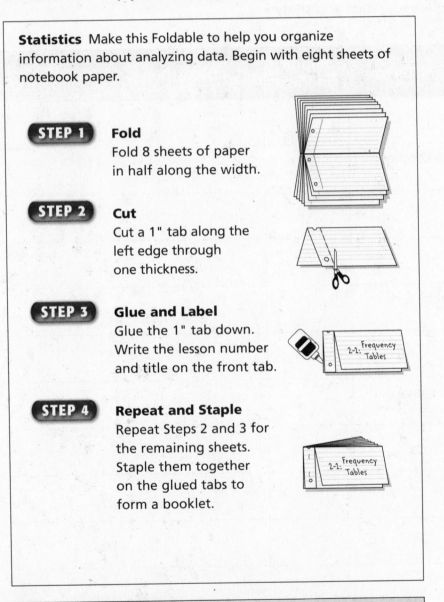

**STEP 1**   **Fold**
Fold 8 sheets of paper in half along the width.

**STEP 2**   **Cut**
Cut a 1" tab along the left edge through one thickness.

**STEP 3**   **Glue and Label**
Glue the 1" tab down. Write the lesson number and title on the front tab.

2-1: Frequency Tables

**STEP 4**   **Repeat and Staple**
Repeat Steps 2 and 3 for the remaining sheets. Staple them together on the glued tabs to form a booklet.

2-1: Frequency Tables

**NOTE-TAKING TIP:** When you take notes, it is sometimes helpful to make a graph, diagram, picture, chart, or concept map that presents the information introduced in the lesson.

**Chapter 2**

This is an alphabetical list of new vocabulary terms you will learn in Chapter 2. As you complete the study notes for the chapter, you will see Build Your Vocabulary reminders to complete each term's definition or description on these pages. Remember to add the textbook page number in the second column for reference when you study.

| Vocabulary Term | Found on Page | Definition | Description or Example |
|---|---|---|---|
| bar graph | | | |
| box-and-whisker plot | | | |
| cluster | | | |
| data | | | |
| frequency table | | | |
| histogram | | | |
| interquartile range | | | |
| interval | | | |
| leaf | | | |
| line graph | | | |
| line plot | | | |
| lower extreme | | | |

| Vocabulary Term | Found on Page | Definition | Description or Example |
|---|---|---|---|
| lower quartile | | | |
| mean | | | |
| measures of central tendency | | | |
| median | | | |
| mode | | | |
| outlier | | | |
| range | | | |
| scale | | | |
| scatter plot | | | |
| statistics | | | |
| stem | | | |
| stem-and-leaf-plot | | | |
| upper extreme | | | |
| upper quartile | | | |

## 2–1 Frequency Tables

### WHAT YOU'LL LEARN

- Organize and interpret data in a frequency table.

---

**BUILD YOUR VOCABULARY** (pages 32–33)

**Statistics** deals with [ ], organizing, and interpreting [ ].

**Data** are pieces of information which are often numerical.

A **frequency table** shows the number of pieces of [ ] that fall within given [ ].

The **scale** allows you to record all of the data, including the [ ] value and the [ ] value.

The **interval** separates the scale into equal parts.

---

### FOLDABLES

## ORGANIZE IT

Find some data in a newspaper. Under the tab for Lesson 2–1 display the data in a table and draw a frequency table for the data.

> 2-1: Frequency Tables

---

**EXAMPLE** Make a Frequency Table

**① FOOTBALL** Winning Super Bowl scores from 1983 to 2002 are listed below. Make a frequency table of the data.

**Step 1** Choose an appropriate [ ] and scale for the data. The [ ] should include the least value, [ ], and the greatest value, [ ].

| Winning Scores | | | |
|---|---|---|---|
| 20 | 34 | 23 | 34 |
| 31 | 35 | 27 | 49 |
| 30 | 52 | 37 | 20 |
| 55 | 20 | 42 | 39 |
| 46 | 38 | 38 | 27 |

**Source:** superbowl.com

interval: 9
scale: 20 to 55 ⎬ The scale includes all of the data, and the interval separates it into equal parts.

---

**Step 2** Draw a table with three columns and label the columns. *Scores, Tally,* and *Frequency*.

**Step 3** Complete the table.

| Scores | Tally | Frequency |
|--------|-------|-----------|
| 20–28 | HHT I | |
| 29–37 | HHT I | 6 |
| | HHT | 5 |
| 45–55 | | 3 |

**Your Turn** The daily high temperatures for the last two weeks of August in Cleveland, Ohio are listed below. Make a frequency table of the data.

| High Temperature (°F) | | |
|-----|-----|-----|
| 83 | 85 | 76 |
| 88 | 69 | 75 |
| 81 | 90 | 83 |
| 72 | 79 | 81 |
| 83 | 92 | |

**EXAMPLE** Make and Use a Frequency Table

**2** **MUSIC** Kaley asked her classmates about their favorite types of music. The results are shown in the table. Make a frequency table of the data. Then determine the favorite and least favorite types of music.

| Favorite Types of Music | | | | |
|---|---|---|---|---|
| P | R | F | P | F |
| F | R | F | P | P |
| F | P | F | C | P |
| C | J | R | R | F |
| J | R | P | P | F |

R=rock, J=jazz, C=country, F=top 40, P=rap

Draw a table with three columns. In the first column, list the types of music. Then complete the rest of the table.

| Music | Tally | Frequency |
|---|---|---|
| rock | ⊬Ⅱ | |
| | \|\| | 2 |
| country | ⊬Ⅱ \|\|\| | 2 |
| top 40 | | 8 |
| | ⊬Ⅱ \|\|\| | 8 |

The two favorite types were ☐ and rap, with ☐ tallies. The least favorite types were ☐ and country, with ☐ tallies.

**Your Turn** Samantha asked her classmates about their favorite colors. The results are shown in the table. Make a frequency table of the data. Then determine the favorite and least favorite colors.

| Favorite Colors | | | | | | | | | |
|---|---|---|---|---|---|---|---|---|---|
| Y | R | B | P | L | R | P | B | Y | R |
| P | Y | R | L | Y | Y | P | Y | B | Y |

Y=yellow, R=red, B=blue, P=pink, L=purple

**EXAMPLE** Interpret Data

**3 TEMPERATURE** The frequency table shows the record high temperatures reported by each state of the United States. How many states have reported temperatures above 111°F?

| Temp (°F) | Tally | Frequency |
|---|---|---|
| 100–105 | $\cancel{||||}$ | |
| 106–111 | $\cancel{||||}$ $\cancel{||||}$ || | 12 |
| 112–117 | $\cancel{||||}$ $\cancel{||||}$ $\cancel{||||}$ | | |
| | $\cancel{||||}$ $\cancel{||||}$ |||| | 14 |
| 124–129 | | 2 |
| | | | 1 |

There are four categories with temperatures above 111°F.

So, ⬜ + 14 + ⬜ + 1 or ⬜ of the states

reported temperatures ⬜ 111°F.

**Your Turn** The frequency table shows the results of a survey of the weights of boys in a seventh grade class. How many of the boys weigh less than 100 pounds?

| Weight | Tally | Frequency |
|---|---|---|
| 70–79 | ||| | 3 |
| 80–89 | $\cancel{||||}$ | 5 |
| 90–99 | $\cancel{||||}$ || | 7 |
| 100–109 | $\cancel{||||}$ ||| | 8 |
| 110–119 | |||| | 4 |
| 120–129 | | | 1 |

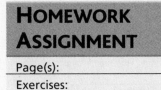

HOMEWORK ASSIGNMENT

Page(s): _____
Exercises: _____

# Making Predictions

**Line graphs** can be useful in predicting [          ] events

when they show trends over [          ] .

## WHAT YOU'LL LEARN

- Make predictions from graphs.

## FOLDABLES

### ORGANIZE IT

Under the tab for Lesson 2-2, include an example of a line graph and explain how it can be used to make predictions.

2-1: Frequency Tables

**EXAMPLE** Use a Line Graph to Predict

**1** TYPING  Enrique is writing a 600-word paper for class. The table shows the time it has taken Enrique to type the paper so far. Make a line graph and predict the total time it will take him to type his paper.

| Time (min) | Words Typed |
|:---:|:---:|
| 0 | 0 |
| 1 | 40 |
| 2 | 85 |
| 3 | 128 |
| 4 | 169 |
| 5 | 214 |
| 6 | 258 |

By looking at the pattern in the graph, you can predict that it will take Enrique

about [          ] minutes to

type his 600-word paper.

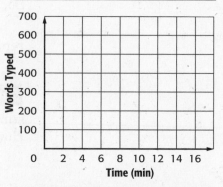

**Your Turn**  During a recent road trip, Helen kept track of the number of miles traveled after each hour of travel time was completed. The table shows her information. Make a line graph and predict how far Helen will travel in 12 hours of travel time.

| Travel Time (hours) | Miles |
|:---:|:---:|
| 0 | 0 |
| 1 | 52 |
| 2 | 110 |
| 3 | 171 |
| 4 | 225 |
| 5 | 290 |
| 6 | 348 |

## WRITE IT

Explain how a line graph can help you to make a prediction.

_____

_____

_____

_____

_____

### BUILD YOUR VOCABULARY (page 33)

A **scatter plot** displays two sets of data on the same

graph and are also useful in making [_____].

**EXAMPLE** Use a Scatter Plot to Predict

**2** **POLLUTION** The scatter plot shows the number of days that San Bernardino, California, failed to meet air quality standards from 1990 to 1998. Use it to predict the number of days of bad air quality in 2004.

By looking at the pattern, you can predict that the number of days of bad air quality in 2004

will be about [_____] days.

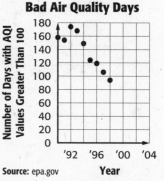

**Bad Air Quality Days**

Source: epa.gov

**Your Turn** Use the scatter plot to predict the gas mileage for a car weighing 5500 pounds.

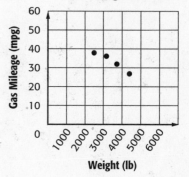

**Gas Mileage of Cars**

## HOMEWORK ASSIGNMENT

Page(s):

Exercises:

_____

_____

## WHAT YOU'LL LEARN

• Construct and interpret line plots.

---

**BUILD YOUR VOCABULARY** (pages 32–33)

A line plot is a diagram that shows the **frequency** of data on a number line.

Data that is grouped closely together is called a **cluster**.

**Outliers** are numbers that are separated from the rest of the data in a data set.

---

**FOLDABLES**™

## ORGANIZE IT

Write a set of data that could be displayed in a line plot. Under the lab for Lesson 2-3, display the data in a line plot.

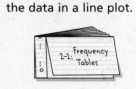

2-1. Frequency Tables

---

**EXAMPLE**   Make a Line Plot

**1** **PRESIDENTS** The table below shows the ages of the U.S. presidents at the time of their inaugurations. Make a line plot of the data.

| Age at Inauguration |
|---|
| 57  51  54  56  61  61  49  49  55  52  57  64  50  51  69 |
| 57  50  47  54  64  58  48  55  51  46  57  65  55  60  54 |
| 61  52  54  62  68  54  56  42  43  46  51  55  56 |

**Step 1**   Draw a number line. Use a scale of 40 to 70 and an interval of 5.

**Step 2**   Place an ✕ above the number that represents the age of each U.S. president.

```
←┼─┼─┼─┼─┼─┼─┼─┼─┼─┼─┼─┼─┼─┼─┼─┼─┼─┼─┼─┼─┼─┼─┼─┼─┼─┼─┼─┼─┼─┼─→
  40      45      50      55      60      65      70
```

**Your Turn**   Make a line plot of the data shown at the right.

| Minutes Studying | | | |
|---|---|---|---|
| 36 | 42 | 60 | 35 |
| 70 | 48 | 55 | 32 |
| 60 | 58 | 42 | 55 |
| 38 | 45 | 60 | 50 |

```
←┼┼┼┼┼┼┼┼┼┼┼┼┼┼┼┼┼┼┼┼┼┼┼┼┼┼┼┼┼┼┼┼┼┼┼┼┼┼┼┼┼┼┼┼┼┼→
   30   35   40   45   50   55   60   65   70   75
```

---

## BUILD YOUR VOCABULARY (page 33)

The **range** is the **difference** between the greatest and least numbers in the data set and is helpful in seeing how spread out the data are.

**EXAMPLES** Use a Line Plot to Analyze Data

**2** **CLIMATE** The line plot shows the number of inches of precipitation that fell in several cities west of the Mississippi River during a recent year. What is the range of the data?

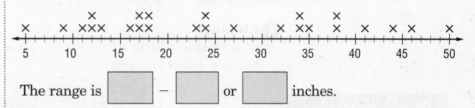

The range is ⬜ – ⬜ or ⬜ inches.

**3** Identify clusters, gaps, and outliners if any exist in the line plot in Example 2 and explain what they mean.

There are data clusters between ⬜ and 13 inches and

between 16 and ⬜ inches. At least half of the data fall

below 25, so most of the selected cities west of the Mississippi River had fewer than 25 inches of precipitation.

**Your Turn** The line plot below shows the ages of students in an introductory computer course at the local community college.

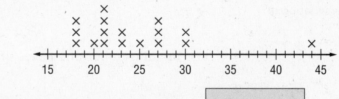

**a.** What is the range of the data? ⬜

**b.** Identify clusters, gaps, and outliers if any exist in the line plot and explain what they mean.

# 2-4 Mean, Median, and Mode

## WHAT YOU'LL LEARN

- Find the mean, median, and mode of a set of data.

**BUILD YOUR VOCABULARY** (page 33)

**Measures of central tendency** can be used to describe the

[        ] of the data.

**EXAMPLE**  Find the Mean

**①ANIMALS** The table below shows the number of species of animals found at 30 major zoos across the United States. Find the mean.

## KEY CONCEPTS

**Measures of Central Tendency**

The **mean** of a set of data is the sum of the data divided by the number of items in the data set.

The **median** of a set of data is the middle number of the ordered data, or the mean of the middle two numbers.

The **mode** or modes of a set of data is the number or numbers that occur most often.

| Number of Species in Major U.S. Zoos | | | | |
|------|------|------|------|------|
| 300 | 400 | 283 | 400 | 175 |
| 617 | 700 | 700 | 715 | 280 |
| 800 | 290 | 350 | 133 | 400 |
| 195 | 347 | 488 | 435 | 640 |
| 232 | 350 | 300 | 300 | 400 |
| 705 | 400 | 800 | 300 | 659 |

**Source:** The World Almanac

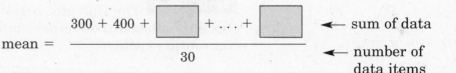

$$\text{mean} = \frac{300 + 400 + \boxed{\phantom{00}} + \ldots + \boxed{\phantom{00}}}{30}$$

← sum of data

← number of data items

The mean number of species of animals is [        ].

**Your Turn** The table below shows the results of a survey of 15 middle school students concerning the number of hours of sleep they typically get each night. Find the mean.

| Nightly Hours of Sleep | | | | |
|---|---|---|---|---|
| 7 | 8 | 6 | 7 | 8 |
| 9 | 5 | 6 | 7 | 7 |
| 8 | 6 | 7 | 8 | 8 |

**EXAMPLE**  Find the Mean, Median, and Mode.

**2** OLYMPICS  The table below shows the number of gold medals won by each country participating in the 2002 Winter Olympic games. Find the mean, median, and mode of the data.

| 2002 Winter Olympics: Gold Medals Won | | | | |
|---|---|---|---|---|
| 12 | 6 | 4 | 3 | 0 |
| 10 | 6 | 4 | 2 | 3 |
| 11 | 2 | 3 | 4 | 2 |
| 1 | 1 | 0 | 2 | 2 |
| 1 | 0 | 0 | 0 | 0 |

**Source:** CBSSportsline.com

mean:    sum of data divided by ⬚, or ⬚

median:  13th number of the ⬚ data, or ⬚

mode:    number appearing ⬚ often, or ⬚

So, the mean, median, and mode are ⬚, ⬚, and

⬚ respectively.

**Your Turn**  The table below shows the number of pets students in an art class at Green Hills Middle School have at home. Find the mean, median, and mode of the data.

| Pets | | | |
|---|---|---|---|
| 0 | 2 | 1 | 0 |
| 1 | 3 | 5 | 2 |
| 0 | 1 | 0 | 2 |
| 3 | 1 | 2 | 0 |

**EXAMPLE** Analyze Data

**③ FIRST FAMILIES** The line plot shows the number of children of United States presidents. Would the mean, median, or mode best represent the number of children?

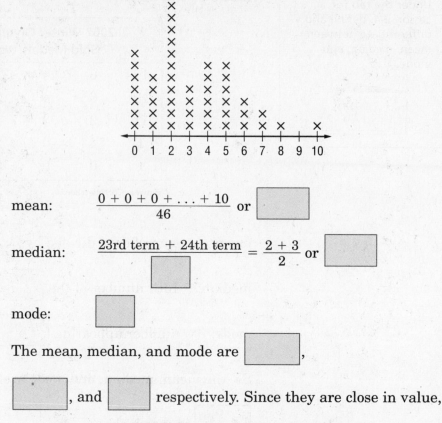

mean: $\dfrac{0 + 0 + 0 + \ldots + 10}{46}$ or [    ]

median: $\dfrac{\text{23rd term} + \text{24th term}}{[\ \ ]} = \dfrac{2 + 3}{2}$ or [    ]

mode: [    ]

The mean, median, and mode are [    ],

[    ], and [    ] respectively. Since they are close in value,

any of the three could be used to represent the data.

**Your Turn** The line plot below shows the number of siblings of each student in a particular classroom. Would the mean, median, or mode best represent the number of siblings?

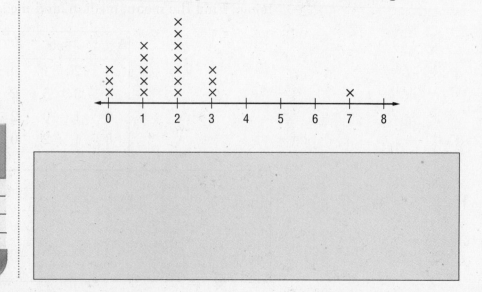

**HOMEWORK ASSIGNMENT**

Page(s): _____

Exercises: _____

# Stem-and-Leaf Plots

## WHAT YOU'LL LEARN

- Construct and interpret stem-and-leaf plots.

**BUILD YOUR VOCABULARY** (pages 32–33)

In a **stem-and-leaf plot**, the data are organized from

[          ] to [          ] .

The digits of the [          ] place value usually form the

**leaves** and the next place value digits form the **stems**.

**FOLDABLES**

## ORGANIZE IT

Under the tab for Lesson 2-5, give an example of a set of data for which a stem-and-leaf plot would be appropriate. Draw the stem-and-leaf plot.

2-1: Frequency Tables

**EXAMPLE**   Construct a Stem-and-Leaf Plot

**1** BASEBALL  The table below shows the number of home runs that Babe Ruth hit during his career from 1914 to 1935. Make a stem-and-leaf plot of the data.

| Home Runs | | | |
|---|---|---|---|
| 0 | 54 | 25 | 46 |
| 4 | 59 | 47 | 41 |
| 3 | 35 | 60 | 34 |
| 2 | 41 | 54 | 6 |
| 11 | 22 | 46 | |
| 29 | 46 | 49 | |

**Source:** baberuth.com

**Step 1**  The digits in the [          ] place value will form the

leaves and the remaining digits will form the

[          ] . In this data, [          ] is the least value,

and [          ] is the greatest. So, the ones digit will

form the [          ] and the [          ] digit will

form the stems.

**Step 2** List the stems 0 to [ ] in order from least to greatest in the *Stem* column. Write the leaves, the [ ] digits of the home runs, to the [ ] of the corresponding stems.

**Step 3** Order the leaves and write a *key* that explains how to read the stems and leaves

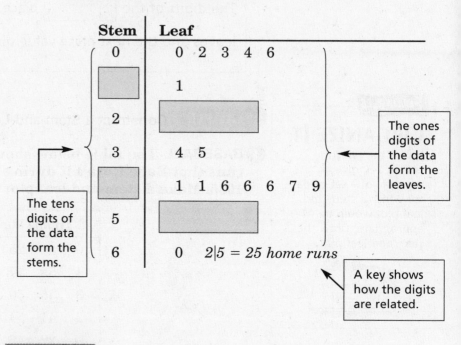

| Stem | Leaf |
|------|------|
| 0 | 0  2  3  4  6 |
| [ ] | 1 |
| 2 | [ ] |
| 3 | 4  5 |
| [ ] | 1  1  6  6  6  7  9 |
| 5 | [ ] |
| 6 | 0      2|5 = 25 home runs |

The tens digits of the data form the stems.

The ones digits of the data form the leaves.

A key shows how the digits are related.

**Your Turn** The table shows the number of hours spent aboard an airplane for a survey of business men and women. Make a stem-and-leaf plot of the data.

| Hours Aboard an Airplane | | | | | | |
|------|------|------|------|------|------|------|
| 4 | 18 | 0 | 23 | 12 | 7 | 9 |
| 35 | 14 | 6 | 11 | 21 | 19 | 6 |
| 15 | 26 | 9 | 0 | 13 | 22 | 10 |

## EXAMPLE   Analyze Data

**2** **FITNESS** The stem-and-leaf plot below shows the number of miles that Megan biked each day during July. Find the range, median, and mode of the data.

| Stem | Leaf |
|------|------|
| 0 | 5  5  5  6 |
| 1 | 0  0  0  0  1  2  2  5  8  8  9 |
| 2 | 1  2  5  8 |
| 3 | 0    2|5 = 25 miles |

range: greatest distance − least distance = ☐ − ☐

or ☐ miles

median: middle value, or ☐ miles

mode: most frequent value, or ☐ miles

**Your Turn** The stem-and-leaf plot below shows the number of inches of snow that fell in Hightown during the month of January for the past 15 years. Find the range, median, and mode.

| Stem | Leaf |
|------|------|
| 0 | 1  3  5  7  9 |
| 1 | 0  0  0  2  4  4  7  8 |
| 2 | 2  6    1|2 = 12 inches |

## EXAMPLE   Make Conclusions About Data

**3** **ANIMALS** The table shows the average life span of several animals. Make a stem-and-leaf plot of the data. Then use it to describe how the data are spread out.

| Animal | Years | Animal | Years | Animal | Years |
|--------|-------|--------|-------|--------|-------|
| Baboon | 20 | Chipmunk | 6 | Guinea Pig | 4 |
| Black Bear | 18 | Cow | 15 | Horse | 20 |
| Polar Bear | 20 | Deer | 8 | Mouse | 3 |
| Camel | 12 | Dog | 12 | Squirrel | 10 |
| Cat | 12 | Elephant | 40 | Tiger | 16 |
| Chimpanzee | 20 | Giraffe | 10 | Zebra | 15 |

**Source:** *The World Almanac*

The least value is **3**, and the greatest value is **40.** The tens digit form the stems, and the ones digits form the leaves.

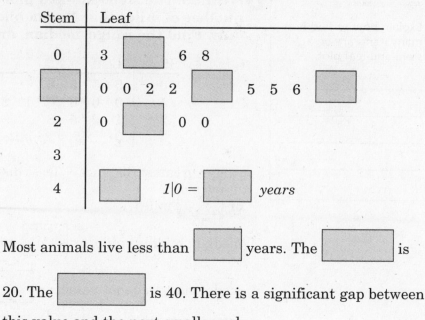

| Stem | Leaf |
|------|------|
| 0 | 3 ☐ 6 8 |
| ☐ | 0 0 2 2 ☐ 5 5 6 ☐ |
| 2 | 0 ☐ 0 0 |
| 3 | |
| 4 | ☐    $1|0 = $ ☐ *years* |

Most animals live less than ☐ years. The ☐ is 20. The ☐ is 40. There is a significant gap between this value and the next smaller value.

**Your Turn** The table below shows the test scores earned by a class of middle school math students on a chapter test. Make a stem-and-leaf plot of the data. Then use it to describe how the data are spread out.

| Test Scores | | | | | |
|----|----|----|----|----|----|
| 82 | 94 | 75 | 85 | 88 | 90 |
| 93 | 80 | 58 | 77 | 86 | 92 |
| 81 | 85 | 94 | 79 | 86 | 93 |
| 87 | 93 | 96 | 80 | 82 | 76 |

**HOMEWORK ASSIGNMENT**

Page(s):

Exercises:

# Box-and-Whisker Plots

**BUILD YOUR VOCABULARY** (pages 32–33)

A **box-and-whisker plot** is a diagram that summarizes

[  ] by dividing it into [  ] parts.

The median of the upper [  ] is called the **upper quartile (UQ)**.

The median of the [  ] half is called the **lower quartile (LQ)**.

**EXAMPLE**   Construct a Box-and-Whisker Plot

**1** NUTRITION  The grams of fat per serving of items from the meat, poultry, and fish food group are shown in the table. Make a box-and-whisker plot of the data.

| Nutrition Facts | | | |
|---|---|---|---|
| **Item** | **Fat (gm)** | **Item** | **Fat (gm)** |
| Bacon | 9 | Ham | 14 |
| Beefsteak | 15 | Pork chop | 19 |
| Bologna | 16 | Roast beef | 5 |
| Crabmeat | 3 | Salmon | 5 |
| Fish sticks | 3 | Sardines | 9 |
| Fried shrimp | 10 | Trout | 9 |
| Ground beef | 18 | Tuna | 7 |

**Source:** *The World Almanac*

**Step 1**   Order the data from least to greatest.

**Step 2**   Find the median and the quartiles.

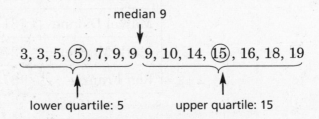

median 9

3, 3, 5, ⑤, 7, 9, 9  9, 10, 14, ⑮, 16, 18, 19

lower quartile: 5          upper quartile: 15

**Step 3** Draw a number line. The scale should include the median, the quartiles, and the lower and upper extremes. Graph the values as points above the line.

**Step 4** Draw the box and whiskers.

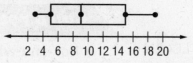

|  |  |  |  |  |  |  |  |  |  |
| 2 | 4 | 6 | 8 | 10 | 12 | 14 | 16 | 18 | 20 |

**REMEMBER IT**

The first step in making a box-and-whisker plot is to order the data.

**Your Turn** The number of students attending class each day are shown in the table. Make a box-and-whisker plot of the data.

| Attendance |    |    |    |    |
|----|----|----|----|----|
| 16 | 19 | 20 | 19 | 18 |
| 19 | 20 | 20 | 17 | 19 |
| 20 | 20 | 19 | 18 | 15 |

**BUILD YOUR VOCABULARY** (pages 32–33)

The least value on a [        ] is called the **lower extreme**.

The [        ] value on a scale is called the **upper extreme**.

**EXAMPLE** Analyze Data

**2 HOCKEY** The table shows the ten all-time leading scorers in the National Hockey League through a recent season. Make a box-and-whisker plot of the data. Then use it to describe how the data are spread.

| NHL Leading Scorers |  |  |  |
|----|----|----|----|
| Player | Goals | Player | Goals |
| Wayne Gretzky | 894 | Steve Yzerman | 645 |
| Gordie Howe | 801 | Phil Esposito | 717 |
| Marcel Dionne | 731 | Ray Bourque | 410 |
| Mark Messier | 627 | Mario Lemieux | 648 |
| Ron Francis | 487 | Paul Coffey | 396 |

**Source:** *The World Almanac*

Find the median, the quartiles, and the extremes.
Then construct the plot.

median = $\dfrac{645 + 648}{2}$ = ☐

LQ = ☐          lower extreme = ☐

UQ = ☐          upper extreme = ☐

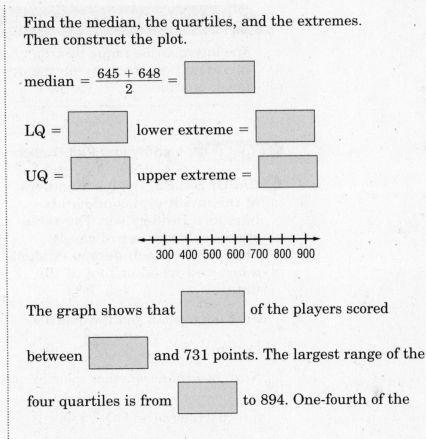

300  400  500  600  700  800  900

The graph shows that ☐ of the players scored

between ☐ and 731 points. The largest range of the

four quartiles is from ☐ to 894. One-fourth of the

players scored within this range.

**Your Turn** The table shows the
commute time from home to
school for fifteen middle school
students. Make a box-and-whisker
plot of the data. Then use it to
describe how the data are spread.

| Commute Time | | | | |
|---|---|---|---|---|
| 25 | 14 | 7 | 18 | 10 |
| 46 | 21 | 5 | 11 | 18 |
| 23 | 17 | 9 | 13 | 12 |

## BUILD YOUR VOCABULARY (page 32)

The **interquartile range** describes how data are spread out. It is the difference between the upper quartile and the lower quartile.

**EXAMPLE** Identify and Plot Outliers

**3 CANDY SALES** Twelve members of the music club sold candy bars as a fund-raiser. The table shows the number of candy bars sold by each person. Make a box-and-whisker plot of the data.

| Candy Sold per Student | | | |
|---|---|---|---|
| 23 | 69 | 27 | 60 |
| 51 | 46 | 81 | 53 |
| 46 | 54 | 39 | 55 |

Find the median and the quartiles.

median = 52          LQ =  42.5          UQ = 57.5

Next, determine whether there are any outliers.

intequartile range: UQ − LQ = ☐ − ☐ or ☐

So, outliers are data more than 1.5(15) or 22.5 from the quartiles. The lower limit is ☐ −22.5 or ☐ . The upper limit is ☐ + 22.5 or ☐ . Any data point that is less than ☐ or greater than ☐ is an outlier. So ☐ is an outlier.

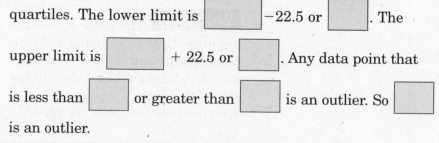

**Your Turn** Make a box-and-whisker plot of the data shown.

| Points Scored | | | |
|---|---|---|---|
| 12 | 14 | 16 | 14 |
| 10 | 12 | 12 | 4 |
| 12 | 14 | 16 | 14 |

**HOMEWORK ASSIGNMENT**

Page(s):

Exercises:

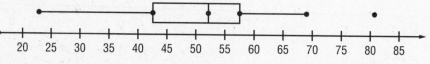

# 2-7 Bar Graphs and Histograms

## WHAT YOU'LL LEARN

- Construct and interpret bar graphs and histograms.

**BUILD YOUR VOCABULARY** (page 32)

A **bar graph** is one method of [ ] data by using solid bars to represent quantities.

**EXAMPLE** Construct a Bar Graph

**① TOURISM** The table below shows the average number of vacation days per year for people in various countries. Make a bar graph to display the data.

| Country | Vacation Days per Year |
|---|---|
| Italy | 42 |
| France | 37 |
| Germany | 35 |
| Brazil | 34 |
| United Kingdom | 28 |
| Canada | 26 |
| Korea | 25 |
| Japan | 25 |
| United States | 13 |

**Source:** *The World Almanac*

**Step 1** Draw and label the axes. Then choose a [ ] on the vertical axis so that it includes all of the vacation days per year.

**Step 2** Draw a [ ] to represent each category.

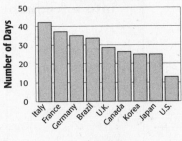

**Vacation Days**

## FOLDABLES

## ORGANIZE IT

Under the tab for Lesson 2-7, draw a sketch of a bar graph and a histogram and describe their similarities and differences.

2-1: Frequency Tables

Your Turn The table shows the average number of miles run each day during training by members of the cross country track team. Make a bar graph to display the data.

| Runner | Miles |
|--------|-------|
| Bob | 9 |
| Tamika | 12 |
| David | 14 |
| Anne | 8 |
| Jonas | 5 |
| Hana | 10 |

## BUILD YOUR VOCABULARY (pages 32)

A **histogram** is a special kind of [          ] graph that uses

bars to represent the frequency of numerical data that

have been organized in [          ].

## WRITE IT

Explain when you would use a bar graph and when you would use a histogram.

_____

_____

_____

_____

**EXAMPLE** Construct a Histogram

2 BASKETBALL The number of wins for the 29 teams of the NBA for the 2000–2001 season have been organized into a frequency table. Make a histogram of the data.

| Number of Wins | Frequency |
|----------------|-----------|
| 11–20 | 3 |
| 21–30 | 4 |
| 31–40 | 4 |
| 41–50 | 10 |
| 51–60 | 8 |

**Step 1** Draw and [____] horizontal and [____]

axes. Add a [____].

**Step 2** Draw a bar to represent the [____] of
each interval.

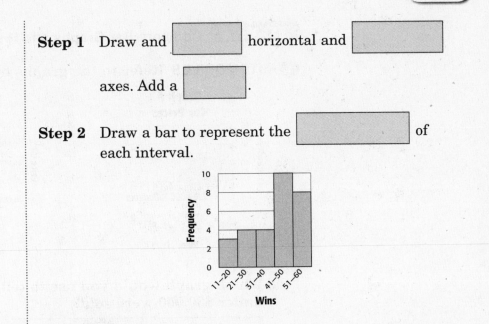

**Your Turn** The speeds of cars on a stretch of interstate are
clocked by a police officer and have been organized into a
frequency table. Make a histogram of the data.

| Speed (mph) | Frequency |
|---|---|
| 50–59 | 2 |
| 60–69 | 14 |
| 70–79 | 18 |
| 80–89 | 3 |

Compare Bar Graphs and Histograms

**3 AUTOMOBILES** Refer to the graphs below.

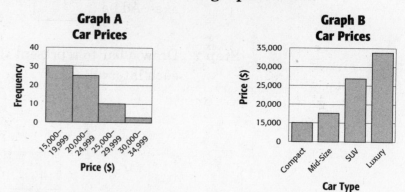

**a. Which graph would you use to tell how many cars under $30,000 were sold?**

**b. Which graph would you use to compare the prices of a mid-size car and an SUV?**

Your Turn Refer to the graphs below.

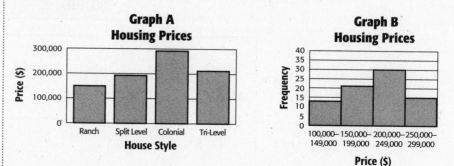

**a. Which graph would you use to tell how many houses sold for $150,000 or greater in a recent year?**

**b. Which graph would you use to compare the price of a ranch style home to the price of a colonial style home?**

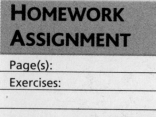

**HOMEWORK ASSIGNMENT**

Page(s):

Exercises:

# Misleading Statistics

**EXAMPLE** Misleading Graphs

**WHAT YOU'LL LEARN**

• Recognize when statistics and graphs are misleading.

**①BUSINESS** The line graphs below show the last 10 weeks of sales for the Crumby Cookie Bakery. Which graph would be better to help convince a bank loan officer to open a $20,000 loan to remodel a kitchen? Why might this graph be considered misleading?

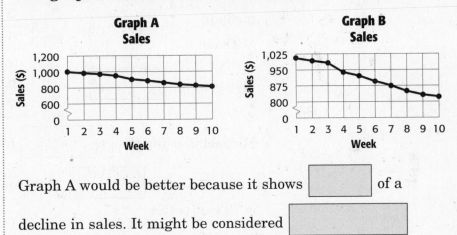

Graph A would be better because it shows [         ] of a

decline in sales. It might be considered [         ]

because it does not show the consistent decline in sales.

**Your Turn** The line graph below shows the monthly profits of a company from May to October. Explain why the graph is misleading.

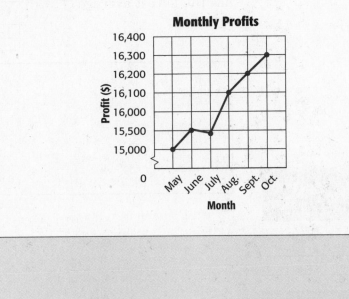

[                                                                    ]

## ORGANIZE IT

Under the tab for Lesson 2-8, explain how to recognize misleading graphs and statistics.

2-1: Frequency Tables

**EXAMPLE** Misleading Statistics

**2 GRADES** Michael and Melissa both claim to be earning a C average, 70% to 79%, in their Latin class. Use the table to explain their reasoning and determine which student is earning a C average.

mean

Michael:

Melissa:

median

Michael:

Melissa:

| Test | Grade (%) | |
|------|-----------|---------|
| | Michael | Melissa |
| 1 | 80 | 88 |
| 2 | 76 | 83 |
| 3 | 73 | 75 |
| 4 | 70 | 70 |
| 5 | 40 | 60 |
| 6 | 25 | 65 |
| 7 | 10 | 62 |

Michael is using the [          ] to describe his grade

rather than the [          ]. Only Melissa's mean or average

is 70% or better.

**Your Turn** Two different grocery stores each claim to have the lowest average prices. Use the table to explain their reasoning and determine which store really has the lowest average prices.

| Item | Store A | Store B |
|------|---------|---------|
| Milk | $1.29 | $1.34 |
| Bread | $1.99 | $1.85 |
| Eggs | $1.19 | $1.09 |
| Soda | $2.29 | $2.99 |
| Coffee | $7.99 | $5.29 |
| Ice Cream | $4.39 | $4.19 |

## HOMEWORK ASSIGNMENT

Page(s):

Exercises:

# BRINGING IT ALL TOGETHER

## STUDY GUIDE

| **FOLDABLES™** | **VOCABULARY PUZZLEMAKER** | **BUILD YOUR VOCABULARY** |
|---|---|---|
| Use your **Chapter 2 Foldable** to help you study for your chapter test. | To make a crossword puzzle, word search, or jumble puzzle of the vocabulary words in Chapter 2, go to: www.glencoe.com/sec/math/t_resources/free/index.php | You can use your completed **Vocabulary Builder** (pages 32–33) to help you solve the puzzle. |

### 2-1
### Frequency Tables

1. Complete the frequency table.

2. Give the interval.

3. Give the scale.

4. How many scores were less than 25?

| Score | Tally | Frequency |
|---|---|---|
| 1–12 | \|\|\|\| | |
| 13–24 | | 7 |
| | ⊬⊬ ⊬⊬ \|\|\| | |

### 2-2
### Making Predictions

**Complete the sentence.**

5. A line graph can be used to predict a future event when it

   shows a _____ over _____ .

**Refer to the graph shown.**

6. Mark the City Zoo graph to show how to predict the attendance in 2005.

7. If the trend continues, predict the

   attendance in 2005.

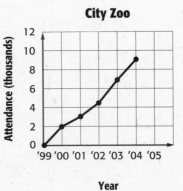

**City Zoo**

### 2-3
### Line Plots

The line plot shows prices for different running shoes.

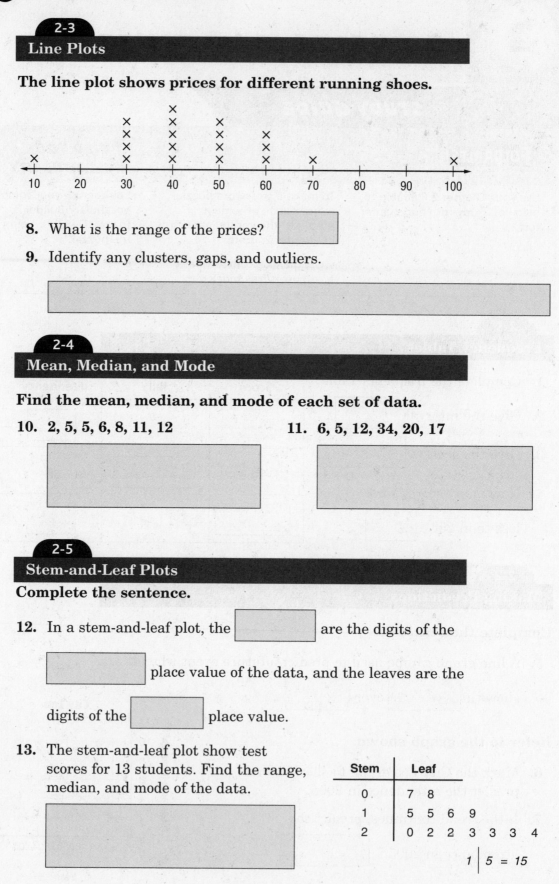

8. What is the range of the prices?

9. Identify any clusters, gaps, and outliers.

### 2-4
### Mean, Median, and Mode

Find the mean, median, and mode of each set of data.

10. 2, 5, 5, 6, 8, 11, 12

11. 6, 5, 12, 34, 20, 17

### 2-5
### Stem-and-Leaf Plots

**Complete the sentence.**

12. In a stem-and-leaf plot, the _____ are the digits of the

_____ place value of the data, and the leaves are the

digits of the _____ place value.

13. The stem-and-leaf plot show test scores for 13 students. Find the range, median, and mode of the data.

| Stem | Leaf |
|------|------|
| 0 | 7 8 |
| 1 | 5 5 6 9 |
| 2 | 0 2 2 3 3 3 4 |

1 | 5 = 15

**2-6**

## Box-and-Whisker Plots

**Complete the construction of a box-and-whisker plot for the data set
7, 10, 11, 13, 13, 15, 18, 19, 20, 20, 31.**

**14.** The median is [ ].

**15.** The lower quartile is [ ].

**16.** The upper quartile is [ ].

**17.** The interquartile range is [ ].

**18.** The outliers are [ ].

**19.** The lower extreme is [ ].

**20.** The upper extreme is [ ].

**21.** Draw a box-and-whisker plot of the data.

**2-7**

## Bar Graphs and Histograms

**Write *true* or *false* for each statement. If the statement is false, replace
the underlined words with words that will make the statement true.**

**22.** A bar graph is used to compare data.

**23.** A histogram shows categories on one of the axes.

**2-8**

## Misleading Statistics

**The table lists the number of wrong answers a student
had on her homework papers this year.**

**24.** Which measure of central tendency might she
use to emphasize her good work? [ ]

| Wrong Answers | | | | |
|---|---|---|---|---|
| 1 | 8 | 2 | 7 | 2 |
| 6 | 8 | 7 | 2 | 4 |
| 7 | 2 | 5 | 8 | 6 |

**25.** Which measure of central tendency best represents her work? Explain.

# ARE YOU READY FOR THE CHAPTER TEST?

**Math Online**

Visit **msmath2.net** to access your textbook, more examples, self-check quizzes, and practice tests to help you study the concepts in Chapter 2.

Check the one that applies. Suggestions to help you study are given with each item.

☐ **I completed the review of all or most lessons without using my notes or asking for help.**

- You are probably ready for the Chapter Test.
- You may want to take the Chapter 2 Practice Test on page 99 of your textbook as a final check.

☐ **I used my Foldables or Study Notebook to complete the review of all or most lessons.**

- You should complete the Chapter 2 Study Guide and Review on pages 96–98 of your textbook.
- If you are unsure of any concepts or skills, refer back to the specific lesson(s).
- You may want to take the Chapter 2 Practice Test on page 99 of your textbook.

☐ **I asked for help from someone else to complete the review of all or most lessons.**

- You should review the examples and concepts in your Study Notebook and Chapter 2 Foldables.
- Then complete the Chapter 2 Study Guide and Review on pages 96–98 of your textbook.
- If you are unsure of any concepts or skills, refer back to the specific lesson(s).
- You may also want to take the Chapter 2 Practice Test on page 99 of your textbook.

Student Signature

Parent/Guardian Signature

Teacher Signature

# CHAPTER 3

# Algebra: Integers

**FOLDABLES™** Use the instructions below to make a Foldable to help you organize your notes as you study the chapter. You will see Foldable reminders in the margin of this Interactive Study Notebook to help you in taking notes.

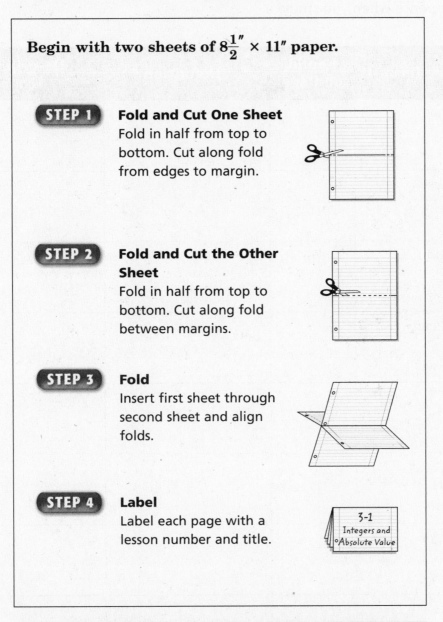

Begin with two sheets of $8\frac{1}{2}''$ × 11″ paper.

**STEP 1** **Fold and Cut One Sheet**
Fold in half from top to bottom. Cut along fold from edges to margin.

**STEP 2** **Fold and Cut the Other Sheet**
Fold in half from top to bottom. Cut along fold between margins.

**STEP 3** **Fold**
Insert first sheet through second sheet and align folds.

**STEP 4** **Label**
Label each page with a lesson number and title.

3-1
Integers and
Absolute Value

**NOTE-TAKING TIPS:** When you take notes, it is helpful to list ways in which the subject matter relates to daily life.

Chapter 3

*Mathematics: Applications and Concepts, Course 2* **61**

# CHAPTER 3

## BUILD YOUR VOCABULARY

This is an alphabetical list of new vocabulary terms you will learn in Chapter 3. As you complete the study notes for the chapter, you will see Build Your Vocabulary reminders to complete each term's definition or description on these pages. Remember to add the textbook page number in the second column for reference when you study.

| Vocabulary Term | Found on Page | Definition | Description or Example |
|---|---|---|---|
| absolute value | | | |
| additive inverse | | | |
| coordinate grid | | | |
| coordinate plane | | | |
| graph | | | |
| integer [IHN-tih-juhr] | | | |
| negative integer | | | |
| opposite | | | |

| Vocabulary Term | Found on Page | Definition | Description or Example |
|---|---|---|---|
| ordered pair | | | |
| origin | | | |
| positive integer | | | |
| quadrant | | | |
| *x*-axis | | | |
| *x*-coordinate | | | |
| *y*-axis | | | |
| *y*-coordinate | | | |

# 3–1 Integers and Absolute Value

## WHAT YOU'LL LEARN

- Read and write integers, and find the absolute value of an integer.

### BUILD YOUR VOCABULARY (pages 62–63)

An **integer** is any _____ from the set {. . ., −4,

_____, −2, −1, 0, 1, _____, 3, 4, . . .}.

To **graph** a _____ on the number line, draw a point on

the line at its _____.

**Negative integers** are integers _____ than zero.

**Positive integers** are integers _____ than zero.

### FOLDABLES

## ORGANIZE IT

Under Lesson 3-1 in your notes, draw a number line and graph a few positive and negative integers. Then write a few real word situations that can be described by negative numbers.

> 3-1
> Integers and
> Absolute Value

### EXAMPLES  Write Integers for Real-Life Situations

Write an integer for each situation.

**1** The seasonal snowfall for Barrow, Alaska was 3 inches above normal.

Because it represents _____ normal, the integer

is _____ .

**2** The total rainfall this month is 2 inches below normal.

Because it represents below normal, the integer is _____ .

**Your Turn**  Write an integer for each situation.

a. The average monthly temperature in Cleveland, Ohio for the month of January was 4 degrees below normal.

b. The total snowfall this month was 5 inches above normal.

The numbers [ ] and 5 are the same [ ] from 0, so −5 and 5 have the same **absolute value**.

**EXAMPLES** Evaluate Expressions

**KEY CONCEPT**

**Absolute Value** The absolute value of an integer is the distance the number is from zero on a number line.

**3** Evaluate the expression |−5|.

On the number line, the graph of −5 is 5 units from 0.

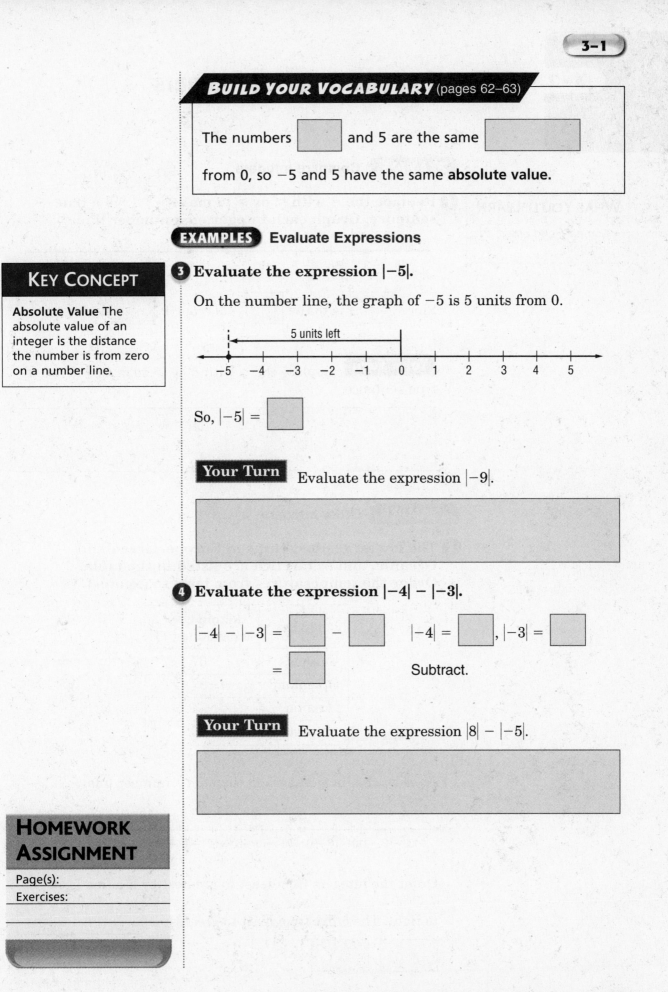

So, |−5| = [ ]

**Your Turn** Evaluate the expression |−9|.

**4** Evaluate the expression |−4| − |−3|.

|−4| − |−3| = [ ] − [ ]     |−4| = [ ] , |−3| = [ ]

= [ ]     Subtract.

**Your Turn** Evaluate the expression |8| − |−5|.

**HOMEWORK ASSIGNMENT**

Page(s):

Exercises:

# Comparing and Ordering Integers

**EXAMPLE** Compare Integers

**WHAT YOU'LL LEARN**

- Compare and order integers

**1** Replace the ● with < or > to make −9 ● −5 a true sentence. Graph each integer on a number line.

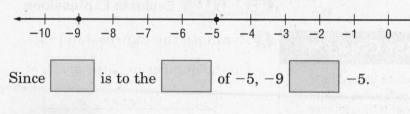

Since [    ] is to the [    ] of −5, −9 [    ] −5.

**Your Turn** Replace the ● with < or > to make −3 ● 6 a true sentence.

**EXAMPLE** Order Integers

**2** The lowest temperatures in Europe, Greenland, Oceania, and Antarctica are listed in the table. Order the temperatures from least to greatest.

| Continent | Record Low Temerature (°F) |
|-----------|----------------------------|
| Europe | −67 |
| Greenland | −87 |
| Oceania | −4 |
| Antarctica | −129 |

**Source:** *The World Almanac*

To order the integers, graph them on a number line.

Order the integers from least to greatest by reading from left to right. The order from least to greatest is [    ], [    ], [    ], [    ].

## ORGANIZE IT

Under Lesson 3-2 in your Foldable, explain how to compare integers. Be sure to include examples.

3-1
Integers and
Absolute Value

**Your Turn** The lowest temperatures on a given day in four cities in the United States are listed in the table. Order the temperatures from least to greatest.

| City | Low Temperature |
|---|---|
| Portland, OR | −12 |
| New York City, NY | −6 |
| Denver, CO | 7 |
| Newport, RI | −3 |

## HOMEWORK ASSIGNMENT

Page(s): _____
Exercises: _____

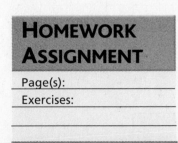

# The Coordinate Plane

## WHAT YOU'LL LEARN

- Graph points on a coordinate plane.

## FOLDABLES

## ORGANIZE IT

Under Lesson 3-3 in your Foldable, record and define key terms about the coordinate system and give examples of each.

3-1
Integers and
Absolute Value

**BUILD YOUR VOCABULARY** (pages 62–63)

A **coordinate plane** is a plane in which a [          ] number line and a vertical number line intersect at their zero points. A coordinate plane is also called a **coordinate grid.**

The [          ] number line of a coordinate plane is called the **x-axis.**

The [          ] number line of a coordinate plane is called the **y-axis.**

The **origin** is the point at which the number lines intersect in a coordinate grid.

An **ordered pair** is a pair of numbers such as (5, −2) used to locate a point in the coordinate plane. The

**x-coordinate** is the [          ] number. The **y-coordinate** is

the [          ] number.

**EXAMPLE**  Name an Ordered Pair

① **Name the ordered pair for point *R* graphed below.**

- Start at the origin.

- Move [          ] to find the x-coordinate of point *R*, which is

[          ].

- Move up to find the

[               ], which is [     ].

So, the ordered pair for point *R* is [          ].

# WRITE IT

When no numbers are shown on the *x*- or *y*-axis, how long is each interval?

_____

_____

_____

_____

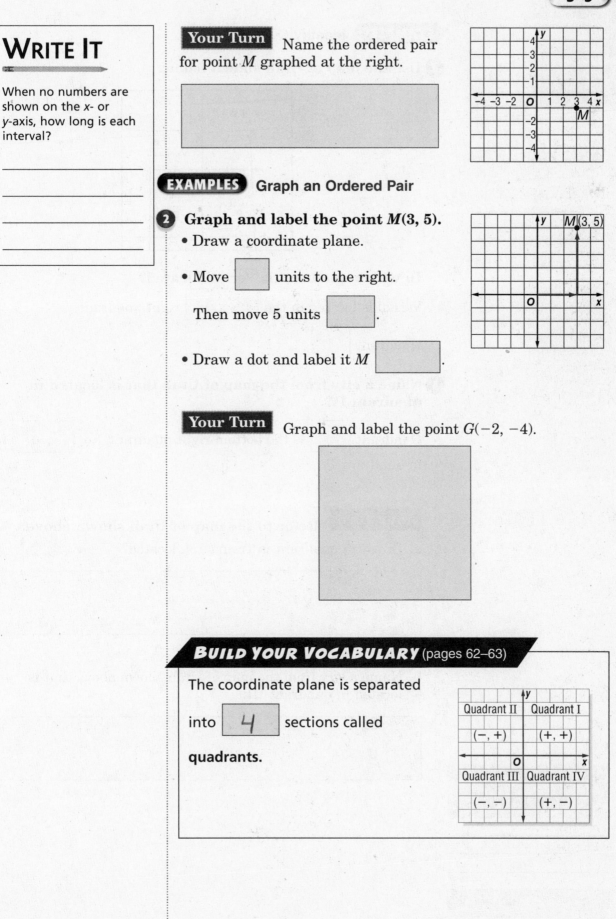

**Your Turn** Name the ordered pair for point *M* graphed at the right.

**EXAMPLES** Graph an Ordered Pair

2 **Graph and label the point *M*(3, 5).**

• Draw a coordinate plane.

• Move [ ] units to the right.

Then move 5 units [ ].

• Draw a dot and label it *M* [ ].

**Your Turn** Graph and label the point *G*(−2, −4).

**BUILD YOUR VOCABULARY** (pages 62–63)

The coordinate plane is separated into 4 sections called **quadrants.**

**EXAMPLES** Identify Quadrants

**3** Use the map of Utah shown below.

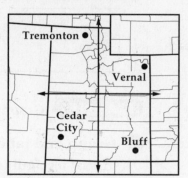

**In which quadrant is Vernal located?**

Vernal is located in the ⬜ right quadrant.

Quadrant ⬜ .

**4** Name a city from the map of Utah that is located in quadrant IV.

Quadrant ⬜ is the bottom-right quadrant. So, ⬜ is in quadrant IV.

**Your Turn** Refer to the map of Utah shown above.

**a.** In which quadrant is Tremonton located?

**b.** Name a city from the map of Utah shown above that is located in quadrant III.

**HOMEWORK ASSIGNMENT**

Page(s):

Exercises:

# 3–4 Adding Integers

**EXAMPLE** Add Integers with the Same Sign

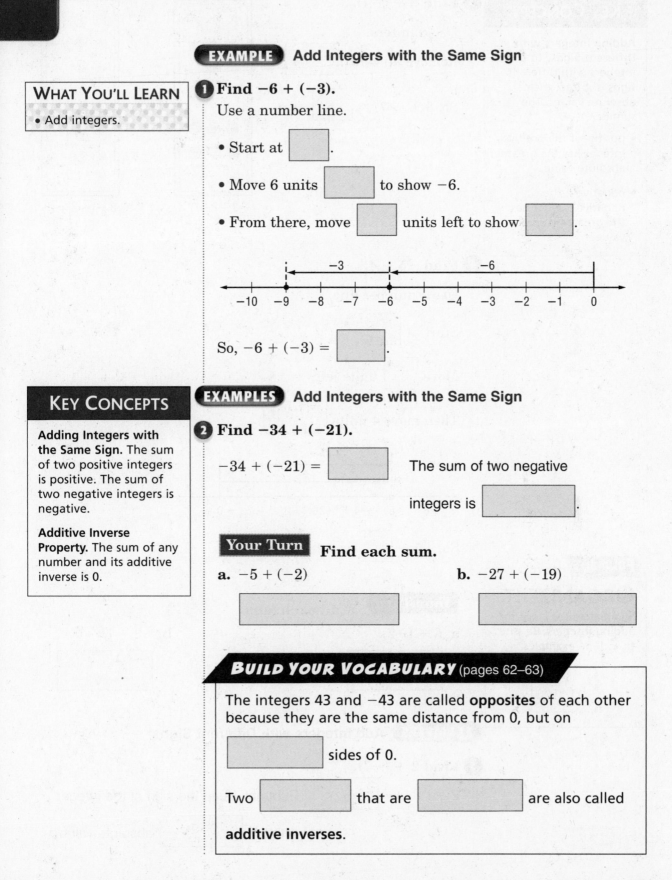

### WHAT YOU'LL LEARN

• Add integers.

**1** **Find −6 + (−3).**
Use a number line.

• Start at [    ].

• Move 6 units [        ] to show −6.

• From there, move [      ] units left to show [      ].

So, −6 + (−3) = [      ].

### KEY CONCEPTS

**Adding Integers with the Same Sign.** The sum of two positive integers is positive. The sum of two negative integers is negative.

**Additive Inverse Property.** The sum of any number and its additive inverse is 0.

**EXAMPLES** Add Integers with the Same Sign

**2** **Find −34 + (−21).**

−34 + (−21) = [      ]     The sum of two negative

integers is [        ].

**Your Turn** Find each sum.

**a.** −5 + (−2)            **b.** −27 + (−19)

**BUILD YOUR VOCABULARY** (pages 62–63)

The integers 43 and −43 are called **opposites** of each other because they are the same distance from 0, but on

[        ] sides of 0.

Two [        ] that are [          ] are also called

**additive inverses**.

**EXAMPLES** Add Integers with Different Signs

### KEY CONCEPT

**Adding Integers with Different Signs.** To add integers with different signs, subtract their absolute values. The sum is:

- positive if the positive integer has the greater absolute value.

- negative if the negative integer has the greater absolute value.

**3** **Find 8 + (−7).**

Use counters.

Remove all zero [        ].

So, 8 + (−7) = [        ].

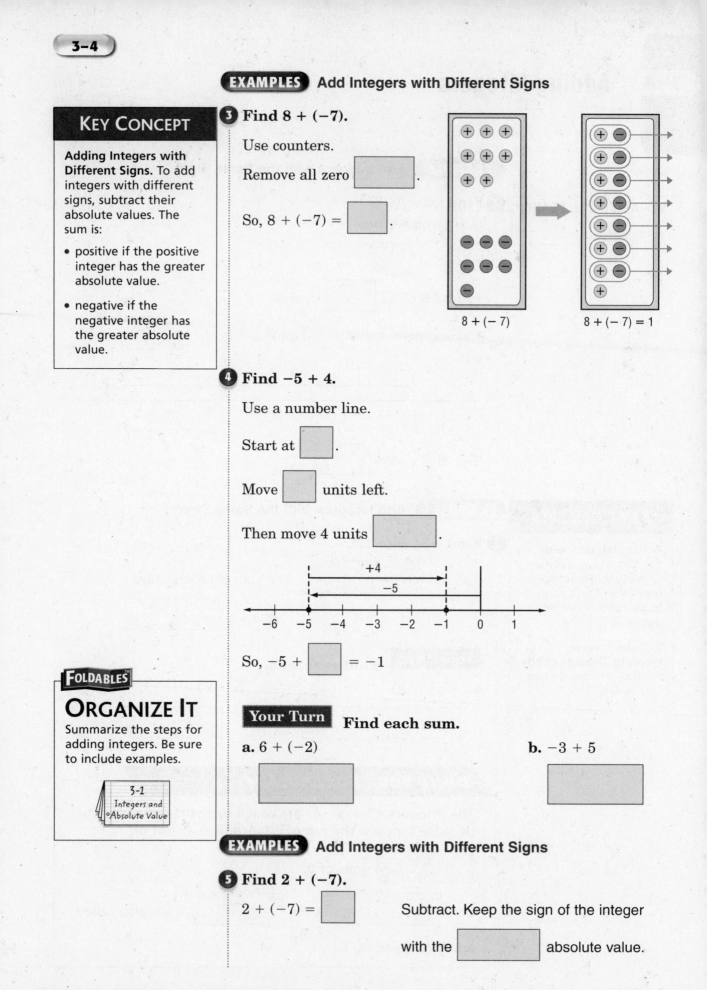

8 + (− 7)

8 + (− 7) = 1

**4** **Find −5 + 4.**

Use a number line.

Start at [        ].

Move [        ] units left.

Then move 4 units [        ].

So, −5 + [        ] = −1

### FOLDABLES™

## ORGANIZE IT

Summarize the steps for adding integers. Be sure to include examples.

> 3-1
> Integers and
> Absolute Value

**Your Turn**  Find each sum.

**a.** 6 + (−2)

**b.** −3 + 5

**EXAMPLES** Add Integers with Different Signs

**5** **Find 2 + (−7).**

2 + (−7) = [        ]

Subtract. Keep the sign of the integer with the [        ] absolute value.

**6** Find −9 + 6.

−9 + 6 = ☐    ☐  . Keep the sign of

the integer with the greater

☐.

**Your Turn** Find each sum.

**a.** 7 + (−3)

☐

**b.** 5 + (−9)

☐

**EXAMPLES** Simplify an Expression with Integers

**7** ALBEGRA **Simplify (−9) + t + 8.**

(−9) + t + 8 = (−9) + 8 + t    Commutative Property of Addition

= ☐ + t    Add.

**REMEMBER IT**

Compare the absolute value of the addends when determining the sign of the sums.

**Your Turn** Simplify 5 + w + (−8).

☐

**HOMEWORK ASSIGNMENT**

Page(s):

Exercises:

# Subtracting Integers

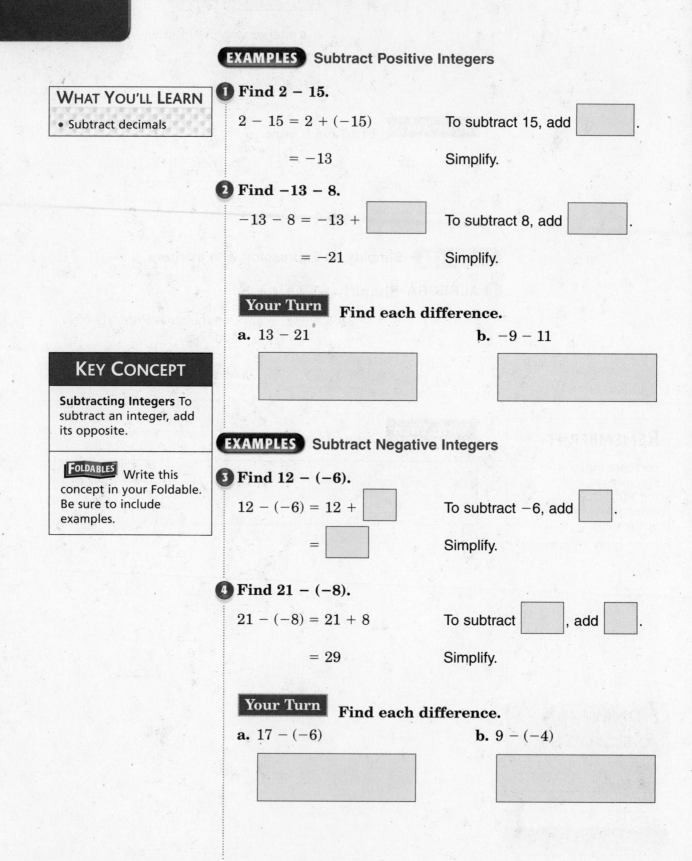

**EXAMPLES** Subtract Positive Integers

**WHAT YOU'LL LEARN**
• Subtract decimals

**1** Find 2 − 15.

$2 - 15 = 2 + (-15)$       To subtract 15, add [ ].

$\phantom{2 - 15} = -13$       Simplify.

**2** Find −13 − 8.

$-13 - 8 = -13 +$ [ ]       To subtract 8, add [ ].

$\phantom{-13 - 8} = -21$       Simplify.

**Your Turn**  Find each difference.

**a.** 13 − 21

**b.** −9 − 11

**KEY CONCEPT**

**Subtracting Integers** To subtract an integer, add its opposite.

**FOLDABLES** Write this concept in your Foldable. Be sure to include examples.

**EXAMPLES** Subtract Negative Integers

**3** Find 12 − (−6).

$12 - (-6) = 12 +$ [ ]       To subtract −6, add [ ].

$\phantom{12 - (-6)} =$ [ ]       Simplify.

**4** Find 21 − (−8).

$21 - (-8) = 21 + 8$       To subtract [ ], add [ ].

$\phantom{21 - (-8)} = 29$       Simplify.

**Your Turn**  Find each difference.

**a.** 17 − (−6)

**b.** 9 − (−4)

**EXAMPLE** Evaluate an Expression

⑤ ALGEBRA Evaluate **g − h** if **g = −2** and **h = −7**.

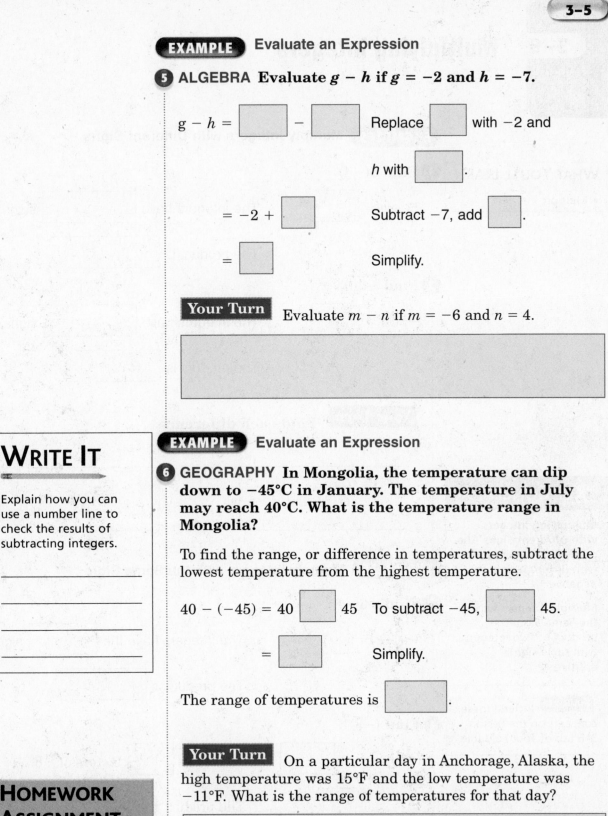

g − h = ☐ − ☐        Replace ☐ with −2 and

h with ☐.

= −2 + ☐        Subtract −7, add ☐.

= ☐        Simplify.

**Your Turn** Evaluate m − n if m = −6 and n = 4.

---

## WRITE IT

Explain how you can use a number line to check the results of subtracting integers.

_____

_____

_____

_____

**EXAMPLE** Evaluate an Expression

⑥ GEOGRAPHY In Mongolia, the temperature can dip down to −45°C in January. The temperature in July may reach 40°C. What is the temperature range in Mongolia?

To find the range, or difference in temperatures, subtract the lowest temperature from the highest temperature.

40 − (−45) = 40 ☐ 45   To subtract −45, ☐ 45.

= ☐        Simplify.

The range of temperatures is ☐ .

**Your Turn** On a particular day in Anchorage, Alaska, the high temperature was 15°F and the low temperature was −11°F. What is the range of temperatures for that day?

---

## HOMEWORK ASSIGNMENT

Page(s):

Exercises:

_____

_____

# Multiplying Integers

**EXAMPLES** Multiply Integers with Different Signs

**WHAT YOU'LL LEARN**

• Multiply integers.

**1** Find 5(−4).

5(−4) = [ ]        The integers have [ ] signs.

The product is [ ].

**2** Find −3(9).

−3(9) = [ ]        The integers have [ ] signs.

The product is [ ].

**Your Turn** Find each difference.

**a.** −5(7)                **b.** 3(−5).

**KEY CONCEPTS**

**Multiplying Integers with Different Signs** The product of two integers with different signs is negative.

**Multiply Integers with the Same Sign** The product of two integers with same sign is positive.

**FOLDABLES** Include these concepts on the Lesson 3–6 tab of your Foldable

**EXAMPLES** Multiply Integers with the Same Sign

**3** Find −6(−8).

−6(−8) = [ ]        The integers have the [ ] sign.

The product is [ ].

**4** Find (−8)².

(−8)² = (−8)[ ]        There are [ ] factors of −8.

= [ ]        The product is [ ].

**Your Turn** Find each product.

**a.** $(-5)^2$

**b.** $-4(-7)$

**EXAMPLES** Simplify and Evaluate Expressions

**5** **ALGEBRA** **Simplify $-5(-7x)$.**

$-5(-7x) = (-5 \cdot (-7))x$      Associative Property of Multiplication

$=$ ⬜      Simplify.

**6** **ALGEBRA** **Evaluate $abc$ if $a = -3$, $b = 5$, and $c = -8$.**

$abc = (-3)(5)(-8)$      Replace ⬜ with $-3$, $b$ with ⬜, and $c$ with ⬜.

$= (-15)(-8)$      Multiply ⬜ and 5.

$=$ ⬜      Multiply $-15$ and $-8$.

**Your Turn**

**a.** Evaluate $xyz$ if $x = -6$, $y = -2$, and $z = 4$.

**b.** Simplify $-6(8x)$.

# Dividing Integers

### WHAT YOU'LL LEARN
• Divide integers.

### KEY CONCEPTS

**Dividing Integers with Different Signs** The quotient of two integers with different signs is negative.

**Dividing Integers with the Same Sign** The quotient of two integers with the same sign is positive.

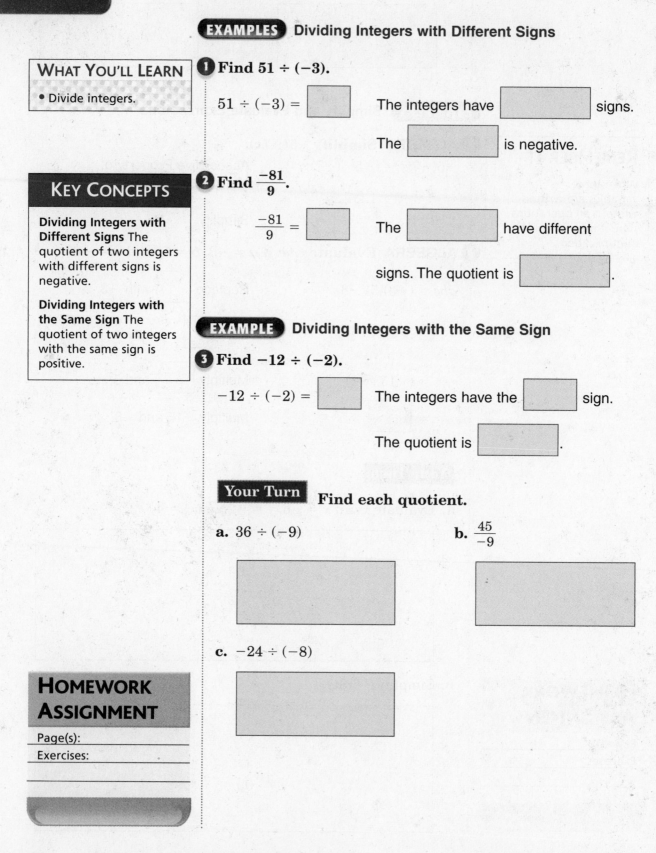

**EXAMPLES** Dividing Integers with Different Signs

**1** Find $51 \div (-3)$.

$51 \div (-3) =$ ☐     The integers have ☐ signs.

The ☐ is negative.

**2** Find $\dfrac{-81}{9}$.

$\dfrac{-81}{9} =$ ☐     The ☐ have different

signs. The quotient is ☐.

**EXAMPLE** Dividing Integers with the Same Sign

**3** Find $-12 \div (-2)$.

$-12 \div (-2) =$ ☐     The integers have the ☐ sign.

The quotient is ☐.

**Your Turn** Find each quotient.

a. $36 \div (-9)$

b. $\dfrac{45}{-9}$

c. $-24 \div (-8)$

### HOMEWORK ASSIGNMENT

Page(s):

Exercises:

# CHAPTER 3

# BRINGING IT ALL TOGETHER

| **FOLDABLES™** | **VOCABULARY PUZZLEMAKER** | **BUILD YOUR VOCABULARY** |
|---|---|---|
| Use your **Chapter 3 Foldable** to help you study for your chapter test. | To make a crossword puzzle, word search, or jumble puzzle of the vocabulary words in Chapter 3, go to: www.glencoe.com/sec/math/ t_resources/free/index.php | You can use your completed **Vocabulary Builder** (*pages 62–63*) to help you solve the puzzle. |

### 3-1
**Integers and Absolute Value**

**Express each of the following in words.**

1. +7

2. −7

3. |7|

4. On the following number line, draw an oval around the *negative* integers and label them negative. Draw a rectangle around the *positive* integers and label them positive.

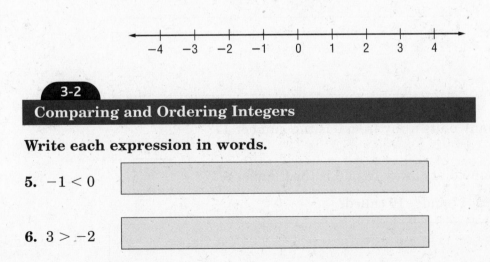

### 3-2
**Comparing and Ordering Integers**

**Write each expression in words.**

5. −1 < 0

6. 3 > −2

*Mathematics: Applications and Concepts, Course 2*     **79**

### 3-3
### The Coordinate Plane

**Look at the coordinate plane at the right. Name the ordered pair for each point graphed.**

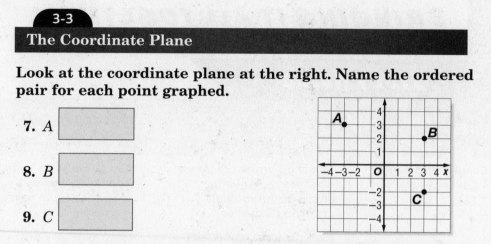

7. *A* 

8. *B* 

9. *C* 

**In the coordinate plane above, identify the quadrant in which each lies.**

10. *A* 

11. *B* 

12. *C* 

### 3-4
### Adding Integers

**Tell how you would solve each of the following on a number line, then add.**

13. $-7 + (-9)$

14. $-7 + 9$

15. How many units away from 0 is the number 17?

16. How many units away from 0 is the number $-17$?

17. What are 17 and $-17$ called?

## 3-5
### Subtracting Integers

**Find each difference. Write an equivalent addition sentence for each.**

**18.** $1 - 5$

**19.** $-2 - 1$

**20.** $-3 - 4$

**21.** $0 - 5$

## 3-6
### Multiplying Integers

**Choose the correct term to complete each sentence.**

**22.** The product of two integers with different signs is (positive, negative).

**23.** The product of two integers with the same sign is (positive, negative).

**Find each product.**

**24.** $(-6)(-4)$     **25.** $-8(5)$     **26.** $-2(3)(-4)$

## 3-7
### Dividing Integers

**Write two division sentences for each of the following multiplication sentences.**

**27.** $6(-3) = 18$

**28.** $-21(-2) = 42$

**29.** $-6(3) = -18$

**30.** $2(-21) = -42$

# CHAPTER 3

**Checklist**

# ARE YOU READY FOR THE CHAPTER TEST?

Visit **msmath2.net** to access your textbook, more examples, self-check quizzes, and practice tests to help you study the concepts in Chapter 3.

**Check the one that applies. Suggestions to help you study are given with each item.**

☐ **I completed the review of all or most lessons without using my notes or asking for help.**

- You are probably ready for the Chapter Test.

- You may want to take the Chapter 3 Practice Test on page 145 of your textbook as a final check.

☐ **I used my Foldables or Study Notebook to complete the review of all or most lessons.**

- You should complete the Chapter 3 Study Guide and Review on pages 142–144 of your textbook.

- If you are unsure of any concepts or skills, refer back to the specific lesson(s).

- You may want to take the Chapter 3 Practice Test on page 145 of your textbook.

☐ **I asked for help from someone else to complete the review of all or most lessons.**

- You should review the examples and concepts in your Study Notebook and Chapter 3 Foldables.

- Then complete the Chapter 3 Study Guide and Review on pages 142–144 of your textbook.

- If you are unsure of any concepts or skills, refer back to the specific lesson(s).

- You may also want to take the Chapter 3 Practice Test on page 145 of your textbook.

Student Signature

Parent/Guardian Signature

Teacher Signature

# Algebra: Linear Equations and Functions

**FOLDABLES** Use the instructions below to make a Foldable to help you organize your notes as you study the chapter. You will see Foldable reminders in the margin of this Interactive Study Notebook to help you in taking notes.

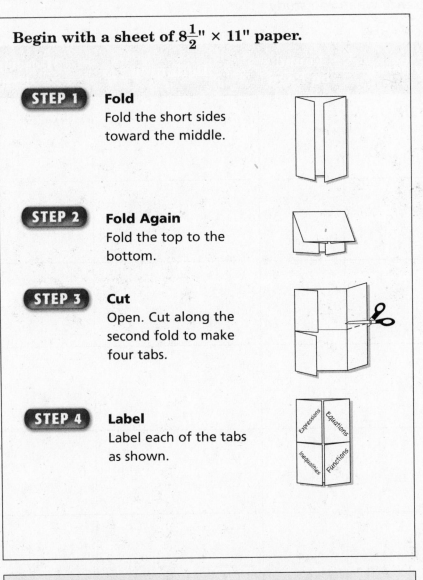

**Begin with a sheet of $8\frac{1}{2}$" × 11" paper.**

**STEP 1** **Fold**
Fold the short sides toward the middle.

**STEP 2** **Fold Again**
Fold the top to the bottom.

**STEP 3** **Cut**
Open. Cut along the second fold to make four tabs.

**STEP 4** **Label**
Label each of the tabs as shown.

*Expressions  Equations  Inequalities  Functions*

 **NOTE-TAKING TIP:** When you take notes, listen or read for main ideas. Then record those ideas in a simplified form for future reference.

## BUILD YOUR VOCABULARY

This is an alphabetical list of new vocabulary terms you will learn in Chapter 4. As you complete the study notes for the chapter, you will see Build Your Vocabulary reminders to complete each term's definition or description on these pages. Remember to add the textbook page number in the second column for reference when you study.

| Vocabulary Term | Found on Page | Definition | Description or Example |
|---|---|---|---|
| Addition Property of Equality | | | |
| Division Property of Equality | | | |
| domain | | | |
| function | | | |
| function table | | | |
| inequality | | | |
| inverse operations | | | |

| Vocabulary Term | Found on Page | Definition | Description or Example |
|---|---|---|---|
| linear equation | | | |
| range | | | |
| slope | | | |
| Subtraction Property of Equality | | | |
| two-step equation | | | |
| work backward strategy | | | |

# ...ressions and Equations

**EXAMPLE** Write a Phrase as an Expression

**①** Write the phrase *twenty dollars less the price of a movie ticket* as an algebraic expression.

**Words** ▼ **Variable** ▼ **Expression**

twenty dollars less the price of a movie ticket

Let ☐ = the price of a movie ticket.

☐

**FOLDABLES**

## ORGANIZE IT
Write two phrases and their algebraic expressions under the **Expressions** tab.

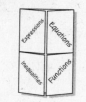

**Your Turn** *five more inches of snow than last year's snowfall*

☐

**EXAMPLES** Write Sentences as Equations

Write each sentence as an algebraic equation.

**②** A number less 4 is 12.

**Words** ▼ **Variable** ▼ **Equation**

A number less 4 is 12.

Let ☐ represent a number.

☐

**③** Twice a number is 18.

**Words** ▼ **Variable** ▼ **Equation**

Twice a number is 18.

Let ☐ represent a number.

☐

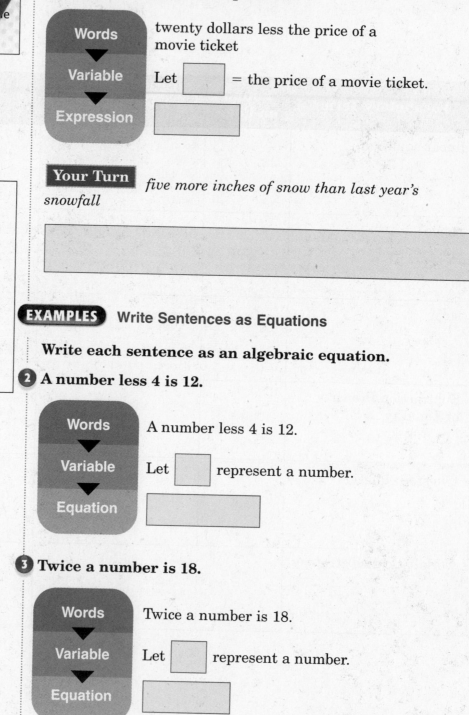

**Your Turn** Write each sentence as an algebraic equation.

**a.** Eight less than a number is 12.

**b.** Four times a number equals 96.

**4** **FOOD** An average American adult drinks more soft drinks than any other beverage each year. Three times the number of gallons of soft drinks plus 27 is equal to the total 183 gallons of beverages consumed. Write the equation that models this situation.

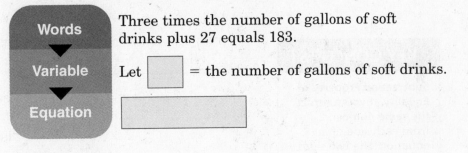

**Words**

**Variable**

**Equation**

Three times the number of gallons of soft drinks plus 27 equals 183.

Let ____ = the number of gallons of soft drinks.

**Your Turn** It is estimated that American adults spend an average of 8 hours per month exercising. This is 26 hours less than twice the number of hours spent watching television each month. Write an equation that models this situation.

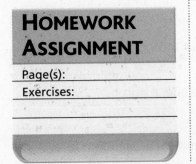

**HOMEWORK ASSIGNMENT**

Page(s): _____

Exercises: _____

# Solving Addition and Subtraction Equations

**WHAT YOU'LL LEARN**

- Solve addition and subtraction equations.

**BUILD YOUR VOCABULARY** (pages 84–85)

Inverse operations, such as [　　　　] 4 from the

equation $x + 4 = 9$, [　　　　] each other.

**EXAMPLE** Solve an Addition Equation

**1** Solve $14 + y = 20$.

$$14 \quad + \quad y \quad = \quad 20$$

Write the equation.

[　　　　]　[　　　　]　　　　[　　　　] 14 from

[　] = [　]

each side. The solution

is [　].

**KEY CONCEPTS**

**Subtraction Property of Equality.** If you subtract the same number from each side of an equation, the two sides remain equal.

**Addition Property of Equality.** If you add the same number from each side of an equation, the two sides remain equal.

**FOLDABLES** Write these properties in your own words under the **Equations** tab.

**Your Turn** Solve $-6 = x + 4$. [　　　　]

**EXAMPLE** Solve a Subtraction Equation

**2** Solve $z - 8 = 12$. Check your solution.

$$z \quad - \quad 8 \quad = \quad 12$$

Write the equation.

[　　　]　[　　　]

Add [　] to each side.

[　] = [　]

Simplify.

Check

$$z - 8 = 12$$

Write the original equation.

[　　] $- 8 = 12$

Replace $z$ with [　].

[　　] $= 12$

Simplify. The solution is [　].

**Your Turn**  Solve $w - 5 = -7$.

---

**EXAMPLE**  Use an Equation to Solve a Problem

**3** SPORTS  If Tiger Woods had scores of $-1$, $-4$, and $-3$ on his first three rounds in another tournament, what would his fourth round score need to be if his final score was $-18$?

**Words** — The sum of the scores for the four rounds was $-18$.

**Variable** — Let $s$ = the score for the fourth round.

**Equation**

scores for the first three rounds plus score for the fourth round equals ☐

$-1 + (-4) + (-3)$ + ☐ = ☐

Simplify.  $-8$ + ☐ = ☐

Add ☐ to each side.  ☐ = ☐

Simplify.  ☐ = ☐

Tiger Woods needs to score ☐ for the fourth round.

**Your Turn**  Kyle wants to hike a trail that is 7 miles long. If he hikes 2, 1, and 2 miles during the first three hours of his hike, how far would he need to hike in the fourth hour in order to complete the trail?

# Solving Multiplication Equations

## WHAT YOU'LL LEARN

- Solve multiplication equations.

## KEY CONCEPTS

**Division Property of Equality.** If you divide each side of an equation by the same nonzero number, the two sides remain equal.

**FOLDABLES** Record the Division Property of Equality in your own words under the Equation tab.

**EXAMPLES** Solving Multiplication Equations

**1** **Solve 39 = 3y. Check your solution.**

$39 = 3y$     Write the equation.

$\boxed{\phantom{00}} = \boxed{\phantom{00}}$     Divide each side of the equation by $\boxed{\phantom{0}}$.

$\boxed{\phantom{00}} = y$     $\boxed{\phantom{00}} \div 3 = \boxed{\phantom{00}}$

Check

$39 = 3y$     Write the equation.

$39 \overset{?}{=} 3\boxed{\phantom{00}}$     Replace $y$ with $\boxed{\phantom{0}}$. Is this sentence true?

$39 = \boxed{\phantom{00}}$

So, the solution is $\boxed{\phantom{00}}$.

**2** **Solve −4z = 60. Check your solution.**

$-4z = 60$     Write the equation.

$\boxed{\phantom{00}} = \boxed{\phantom{00}}$     Divide each side of the equation by $\boxed{\phantom{0}}$.

$z = \boxed{\phantom{00}}$     $60 \div (-4) = \boxed{\phantom{00}}$

Check

$-4z = 60$     Write the equation.

$-4(\boxed{\phantom{00}}) \overset{?}{=} 60$     Replace $z$ with $\boxed{\phantom{0}}$. Is this sentence true?

$\boxed{\phantom{00}} = 60$

So, the solution is $\boxed{\phantom{00}}$.

**Your Turn** Solve each equation. Check your solution.

**a.** $6m = 42$

**b.** $-64 = -16b$

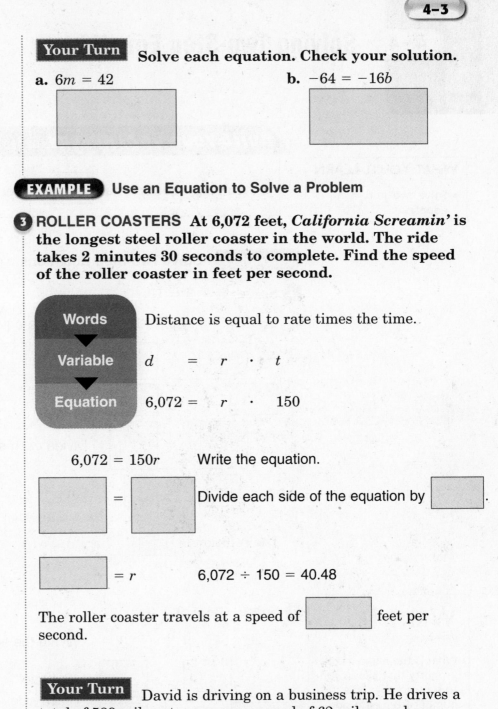

**EXAMPLE** Use an Equation to Solve a Problem

**3** ROLLER COASTERS At 6,072 feet, *California Screamin'* is the longest steel roller coaster in the world. The ride takes 2 minutes 30 seconds to complete. Find the speed of the roller coaster in feet per second.

| Words | Distance is equal to rate times the time. |
|---|---|
| ▼ Variable | $d \quad = \quad r \quad \cdot \quad t$ |
| ▼ Equation | $6{,}072 = \quad r \quad \cdot \quad 150$ |

$6{,}072 = 150r$  Write the equation.

[    ] = [    ]  Divide each side of the equation by [    ].

[    ] $= r$  $6{,}072 \div 150 = 40.48$

The roller coaster travels at a speed of [    ] feet per second.

**Your Turn** David is driving on a business trip. He drives a total of 589 miles at an average speed of 62 miles per hour. How many hours does David spend driving?

## 4–4  Solving Two-Step Equations

### WHAT YOU'LL LEARN

• Solve two-step equations.

**BUILD YOUR VOCABULARY** (pages 84–85)

A **two-step equation** has ☐ different ☐.

**EXAMPLES**  Solve Two-Step Equations

**1** Solve $4x + 3 = 19$.

$4x \quad + \quad 3 \quad = \quad 19$   Write the equation.

☐ $=$ ☐   Subtract ☐ from each side.

☐ $=$ ☐   Simplify.

☐ $=$ ☐   Divide each side by ☐.

☐ $=$ ☐   Simplify.

The solution is ☐.

### WRITE IT

What is the name of the property that allows you to subtract the same number from each side of an equation?

_____

_____

_____

**2** Solve $-3c + 9 = 3$.

$-3c \quad + \quad 9 \quad = \quad 3$   Write the equation.

☐ $=$ ☐   Subtract ☐ from each side.

☐ $=$ ☐   Simplify.

☐ $=$ ☐   Divide each side by ☐.

☐ $=$ ☐   Simplify.

The solution is ☐.

**3** **Solve 6 + 3t = 0.**

| 6 | + | 3t | = | 0 | | Write the equation. |

−6        = −6      [   ]

[   ] = [   ]    Simplify.

[   ] = [   ]    [   ] each side by [   ].

[   ] = [   ]    Simplify.

The solution is [   ].

**Your Turn**   **Solve each equation.**

**a.** $3t - 7 = 14$

**b.** $-8k + 7 = 31$

**c.** $0 = -4x + 32$

**REMEMBER IT**

Always check your solutions by replacing the variable with your answer and simplifying.

**EXAMPLE** Use an Equation to Solve a Problem

④ **PARKS** There are 76,000 acres of state parkland in Georgia. This is 4,000 acres more than three times the number of acres of state parkland in Mississippi. How many acres of state parkland are there in Mississippi?

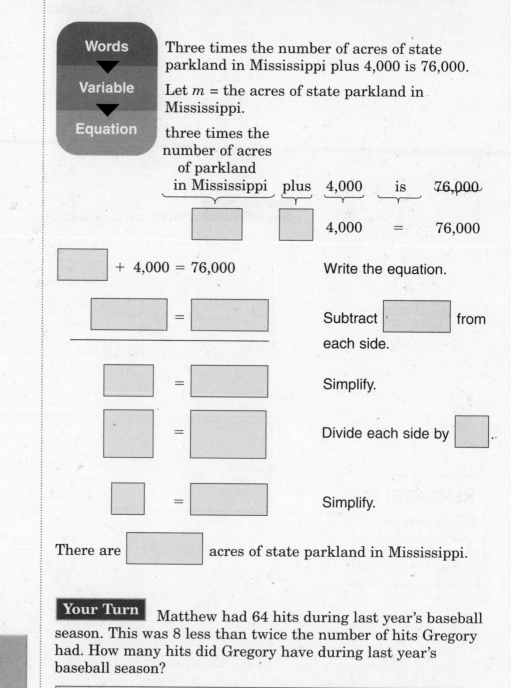

**Words** Three times the number of acres of state parkland in Mississippi plus 4,000 is 76,000.

**Variable** Let $m$ = the acres of state parkland in Mississippi.

**Equation**

three times the number of acres of parkland in Mississippi  plus  4,000  is  76,000

☐ ☐ 4,000 = 76,000

☐ + 4,000 = 76,000     Write the equation.

☐ = ☐     Subtract ☐ from each side.

☐ = ☐     Simplify.

☐ = ☐     Divide each side by ☐.

☐ = ☐     Simplify.

There are ☐ acres of state parkland in Mississippi.

**Your Turn** Matthew had 64 hits during last year's baseball season. This was 8 less than twice the number of hits Gregory had. How many hits did Gregory have during last year's baseball season?

**HOMEWORK ASSIGNMENT**

Page(s):

Exercises:

# Inequalities

## WHAT YOU'LL LEARN

• Solve inequalities.

An **inequality** is a mathematical [          ] that

contains the symbols [      ], >, ≤, or [      ].

## WRITE IT

What is the solution of $x > 5$?

_____

_____

_____

_____

_____

_____

_____

**EXAMPLES** Graphing Solutions of Inequalities

**Graph each inequality on a number line.**

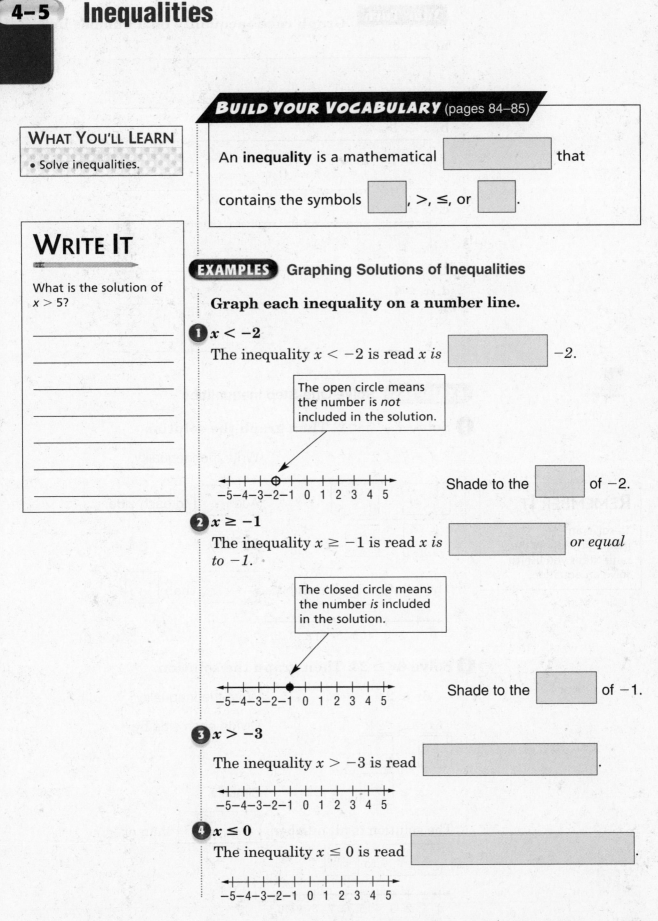

**1** $x < -2$

The inequality $x < -2$ is read $x$ *is* [          ] $-2$.

The open circle means the number is *not* included in the solution.

Shade to the [      ] of $-2$.

**2** $x \geq -1$

The inequality $x \geq -1$ is read $x$ *is* [          ] *or equal to* $-1$.

The closed circle means the number *is* included in the solution.

Shade to the [      ] of $-1$.

**3** $x > -3$

The inequality $x > -3$ is read [                    ].

**4** $x \leq 0$

The inequality $x \leq 0$ is read [                    ].

**Your Turn**  Graph each inequality on a number line.

**a.** $x < 3$

**b.** $x \geq 1$

**c.** $x > 2$

**d.** $x \leq 4$

**EXAMPLES**  Solve One-Step Inequalities

**5** Solve $x - 7 < 2$. Then graph the solution.

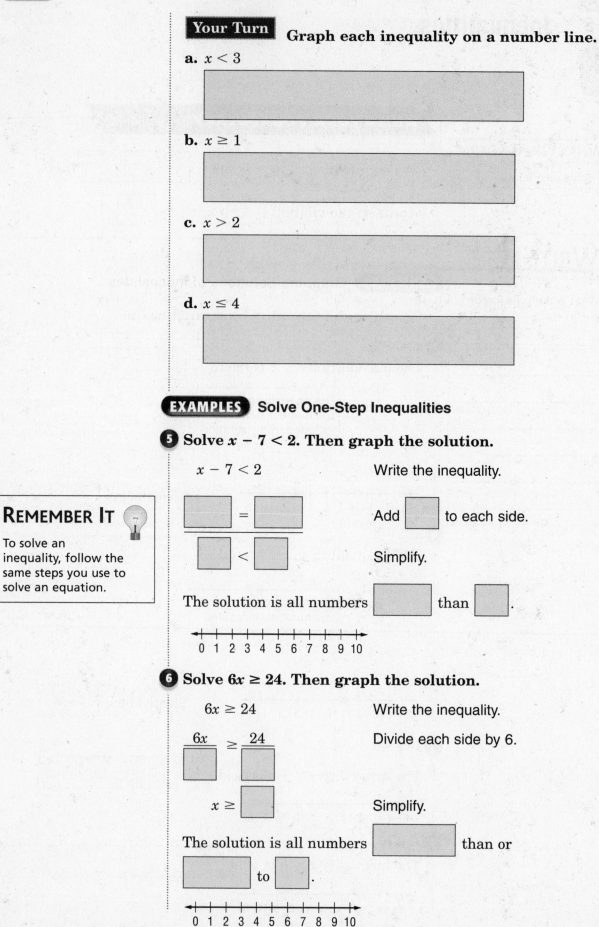

| | |
|---|---|
| $x - 7 < 2$ | Write the inequality. |
| ☐ = ☐ | Add ☐ to each side. |
| ☐ < ☐ | Simplify. |

The solution is all numbers ☐ than ☐.

0  1  2  3  4  5  6  7  8  9  10

**REMEMBER IT**

To solve an inequality, follow the same steps you use to solve an equation.

**6** Solve $6x \geq 24$. Then graph the solution.

| | |
|---|---|
| $6x \geq 24$ | Write the inequality. |
| $\dfrac{6x}{☐} \geq \dfrac{24}{☐}$ | Divide each side by 6. |
| $x \geq ☐$ | Simplify. |

The solution is all numbers ☐ than or ☐ to ☐.

0  1  2  3  4  5  6  7  8  9  10

**Your Turn** Solve each inequality. Graph the solution.

**a.** $x + 3 > 1$

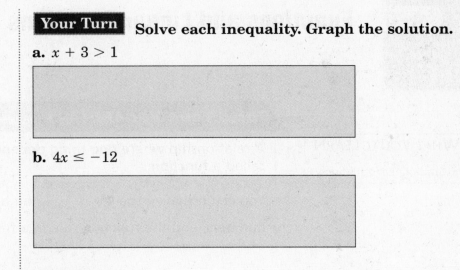

**b.** $4x \le -12$

**EXAMPLE** Use an Inequality to Solve a Problem

**7** **BASEBALL CARDS** Jacob is buying uncirculated baseball cards online. The cards he has chosen are $6.70 each and the Web site charges a $1.50 service charge for each sale. If Jacob has no more than $35 to spend, how many cards can he buy?

Let $c$ represent the number of baseball cards Jacob can buy.

| | | |
|---|---|---|
| $6.70c + 1.50 \le 35.00$ | | Write the equation. |

$$\boxed{\phantom{xx}} = \boxed{\phantom{xx}}$$ 

Subtract $\boxed{\phantom{xx}}$ from each side.

$$\boxed{\phantom{xx}} \le \boxed{\phantom{xx}}$$

Simplify.

$$\boxed{\phantom{xx}} \le \boxed{\phantom{xx}}$$

Divide each side by $\boxed{\phantom{xx}}$.

$c \le 5$       $33.50 \div 6.70 = \boxed{\phantom{xx}}$

Jacob can buy no more than $\boxed{\phantom{xx}}$ baseball cards.

**Your Turn** Danielle is going bowling. The charge for renting shoes is $1.25 and each game costs $2.25. If Danielle has no more than $8 to spend on bowling, how many games can she play?

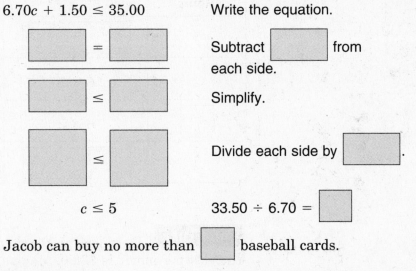

**HOMEWORK ASSIGNMENT**

Page(s):

Exercises:

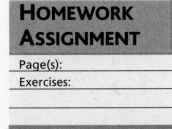

# Functions and Linear Equations

(pages 84–85)

**WHAT YOU'LL LEARN**

• Graph linear equations.

**BUILD YOUR VOCABULARY**

A relationship where one thing depends on another is called a **function**.

You can organize the [　] numbers, [　] numbers, and the function rule in a **function table**.

**REMEMBER IT**

When *x* and *y* are used in an equation, *x* usually represents the input and *y* usually represents the output.

**EXAMPLE** Make Function Table

**1** **WORK** Asha makes $6.00 an hour working at a grocery store. Make a function table that shows Asha's total earnings for working 1, 2, 3, and 4 hours.

| Input | Function | Output |
|---|---|---|
| Number of Hours | Multiply by 6 | Total Earnings ($) |
| 1 | [　] | 6 |
| 2 | 6 × 2 | [　] |
| [　] | 6 × 3 | 18 |
| 4 | [　] | [　] |

**Your Turn** Dave goes to the video store to rent a movie. The cost per movie is $3.50. Make a function table that shows the amount Dave would pay for renting 1, 2, 3, and 4 movies.

The set of input values is called the **domain**.

The set of output values is called the **range**.

An equation like $y = 2a + 1$ is a **linear equation** because

the ⬜ is a ⬜ line.

## WRITE IT

How many points are needed to graph a line? Why is it a good idea to graph more?

_____

_____

_____

_____

**EXAMPLE** Graph Solutions of Linear Equations

**2** Graph $y = x + 3$.

Select any four values for the input $x$. We chose 2, 1, 0, and −1. Substitute these values for $x$ to find the output $y$.

| $x$ | $x + 3$ | $y$ | $(x, y)$ |
|---|---|---|---|
| 2 | ⬜ + 3 | ⬜ | (2, 5) |
| 1 | ⬜ + 3 | 4 | ⬜ |
| 0 | 0 + 3 | ⬜ | ⬜ |
| −1 | ⬜ + 3 | 2 | ⬜ |

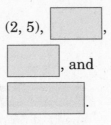

Four solutions are

(2, 5), ⬜ ,

⬜ , and

⬜ .

**Your Turn** Graph $y = 3x - 2$.

**EXAMPLE** Represent Real-World Functions

**3 ANIMALS** Blue whales can reach a speed of 30 miles per hour in bursts when in danger. The equation $d = 30t$ describes the distance $d$ that a whale traveling at that speed can travel in time $t$. Represent this function with a graph.

**Step 1** Select four values for $t$. Select only positive numbers since $t$ represents time. Make a function table.

| $t$ | $30t$ | $d$ | $(t, d)$ |
|---|---|---|---|
| 2 | 30(2) |  | (2, 60) |
| 3 | 30(3) | 90 |  |
| 5 | 30(5) |  |  |
| 6 | 30 | 180 |  |

**Step 2** Graph the ordered pairs and draw a line through the points.

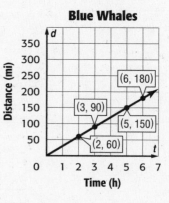

**Blue Whales**

**Your Turn** Susie takes a car trip traveling at an average speed of 55 miles per hour. The equation $d = 55t$ describes the distance $d$ that Susie travels in time $t$. Represent this function with a graph.

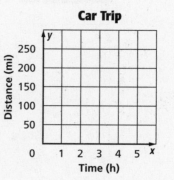

**Car Trip**

**HOMEWORK ASSIGNMENT**

Page(s):

Exercises:

# Lines and Slopes

**WHAT YOU'LL LEARN**

• Find the slope of a line.

**BUILD YOUR VOCABULARY** (pages 84–85)

The change in ⬜ with respect to the change in ⬜ is called the **slope** of a line.

**FOLDABLES**

## ORGANIZE IT

Under the Functions tab, explain how you would use a verbal description, a table of values, an equation or a graph to find the slope of a line.

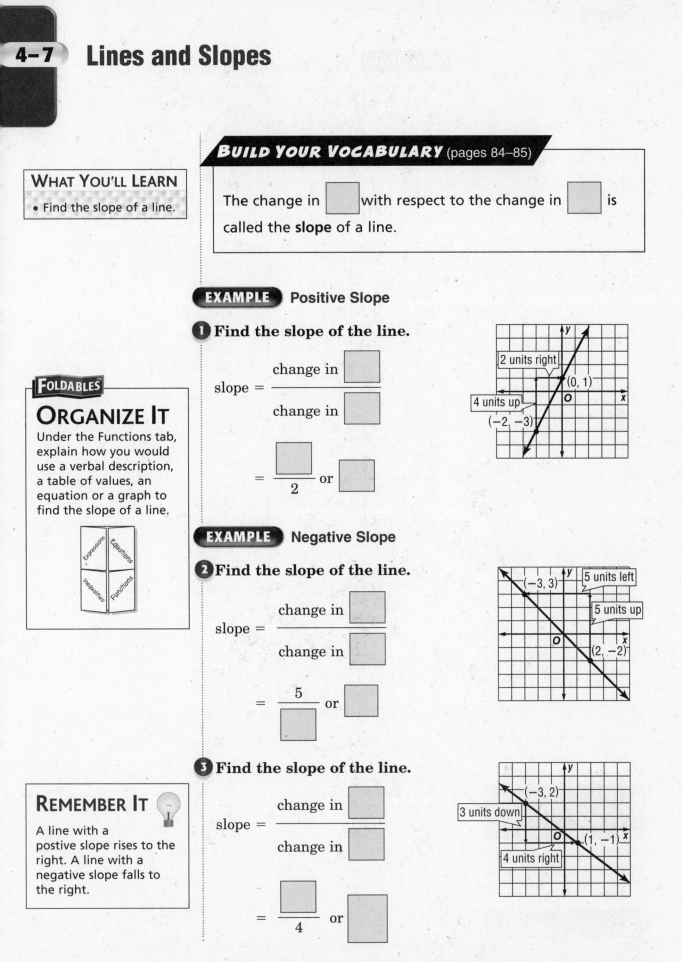

**EXAMPLE** Positive Slope

**1** Find the slope of the line.

$$\text{slope} = \frac{\text{change in } \boxed{\phantom{x}}}{\text{change in } \boxed{\phantom{x}}}$$

$$= \frac{\boxed{\phantom{x}}}{2} \text{ or } \boxed{\phantom{x}}$$

2 units right

(0, 1)

4 units up

(−2, −3)

**EXAMPLE** Negative Slope

**2** Find the slope of the line.

$$\text{slope} = \frac{\text{change in } \boxed{\phantom{x}}}{\text{change in } \boxed{\phantom{x}}}$$

$$= \frac{5}{\boxed{\phantom{x}}} \text{ or } \boxed{\phantom{x}}$$

(−3, 3)    5 units left

5 units up

(2, −2)

**REMEMBER IT**

A line with a postive slope rises to the right. A line with a negative slope falls to the right.

**3** Find the slope of the line.

$$\text{slope} = \frac{\text{change in } \boxed{\phantom{x}}}{\text{change in } \boxed{\phantom{x}}}$$

$$= \frac{\boxed{\phantom{x}}}{4} \text{ or } \boxed{\phantom{x}}$$

(−3, 2)

3 units down

(1, −1)

4 units right

**Your Turn**  **Find the slope of each line.**

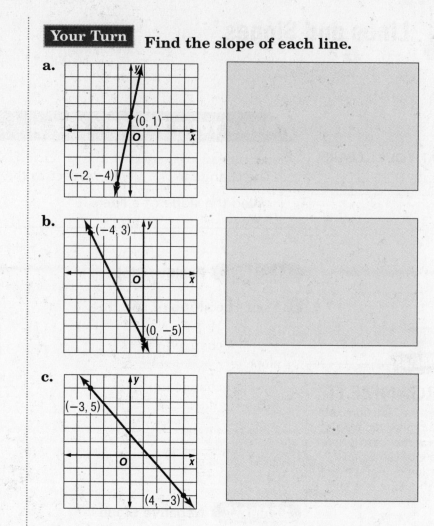

a.

b.

c.

# BRINGING IT ALL TOGETHER

## STUDY GUIDE

| **FOLDABLES**™ | VOCABULARY PUZZLEMAKER | *BUILD YOUR* VOCABULARY |
|---|---|---|
| Use your **Chapter 4 Foldable** to help you study for your chapter test. | To make a crossword puzzle, word search, or jumble puzzle of the vocabulary words in Chapter 4, go to: www.glencoe.com/sec/math/t_resources/free/index.php | You can use your completed **Vocabulary Builder** (*pages 84–85*) to help you solve the puzzle. |

### 4-1
### Writing Expressions and Equations

**Match the phrases with the algebraic expressions that represent them.**

1. seven plus a number

2. seven less a number

3. seven divided by a number

4. seven less than a number

- **a.** $7 - n$
- **b.** $7 \cdot n$
- **c.** $n - 7$
- **d.** $\frac{n}{7}$
- **e.** $7 + n$

**Write each sentence as an algebraic equation.**

5. The product of 4 and a number is 12.

6. Twenty divided by $y$ is equal to $-10$.

### 4-2
### Solving Addition and Subtraction Equations

7. Explain in words how to solve $a - 10 = 3$.

**Solve each equation.**

8. $w + 23 = -11$

9. $35 = z - 15$

**4-3**

## Solving Multiplication Equations

**10.** To solve $-27 = -3d$, divide each side by ⬚ .

**11.** Write and solve an equation that requires you to divide each side by $-2$ in order to solve.

**Solve each equation.**

**12.** $36 = 6k$

**13.** $-7z = 28$

**4-4**

## Solving Two-Step Equations

**14.** Describe in words each step shown for solving $12 + 7s = -9$.

$12 + 7s = -9$

$\dfrac{-12 \qquad\qquad -12}{}$

$7s = -21$

$\dfrac{7s}{7} = \dfrac{-21}{7}$

$s = -3$

**15.** Number the steps in the correct order for solving the equation $-4v + 11 = -5$.

⬚ Simplify

⬚ Write the equation.

⬚ Divide each side by $-4$.

⬚ Simplify.

⬚ Subtract 11 from each side.

⬚ Check the solution.

## 4-5
## Inequalities

**Match each equation with its graph.**

16. $x \le -3$ ☐

17. $x > -3$ ☐

18. $x \ge -3$ ☐

19. $x < -3$ ☐

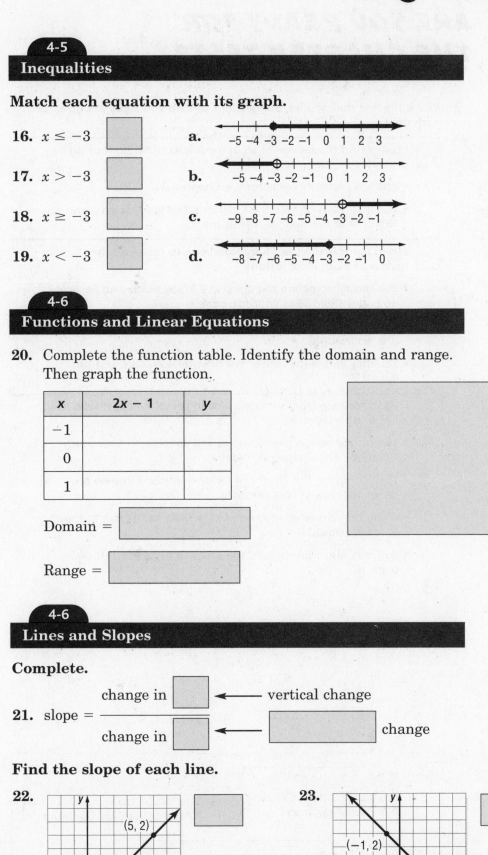

a.
$$-5 \quad -4 \quad -3 \quad -2 \quad -1 \quad 0 \quad 1 \quad 2 \quad 3$$

b.
$$-5 \quad -4 \quad -3 \quad -2 \quad -1 \quad 0 \quad 1 \quad 2 \quad 3$$

c.
$$-9 \quad -8 \quad -7 \quad -6 \quad -5 \quad -4 \quad -3 \quad -2 \quad -1$$

d.
$$-8 \quad -7 \quad -6 \quad -5 \quad -4 \quad -3 \quad -2 \quad -1 \quad 0$$

## 4-6
## Functions and Linear Equations

20. Complete the function table. Identify the domain and range.
Then graph the function.

| x | 2x − 1 | y |
|----|--------|---|
| −1 |        |   |
| 0  |        |   |
| 1  |        |   |

Domain = ☐

Range = ☐

## 4-6
## Lines and Slopes

**Complete.**

21. slope = $\dfrac{\text{change in } \boxed{\phantom{x}} \longleftarrow \text{vertical change}}{\text{change in } \boxed{\phantom{x}} \longleftarrow \boxed{\phantom{xxxxx}} \text{ change}}$

**Find the slope of each line.**

22.
(5, 2)
(2, −1)
☐

23.
(−1, 2)
(2, −1)
☐

# ARE YOU READY FOR THE CHAPTER TEST?

**Math** **Online**

Visit **msmath2.net** to access your textbook, more examples, self-check quizzes, and practice tests to help you study the concepts in Chapter 4.

Check the one that applies. Suggestions to help you study are given with each item.

☐ **I completed the review of all or most lessons without using my notes or asking for help.**

- You are probably ready for the Chapter Test.

- You may want to take the Chapter 4 Practice Test on page 189 of your textbook as a final check.

☐ **I used my Foldables or Study Notebook to complete the review of all or most lessons.**

- You should complete the Chapter 4 Study Guide and Review on pages 186–188 of your textbook.

- If you are unsure of any concepts or skills, refer back to the specific lesson(s).

- You may also want to take the Chapter 4 Practice Test on page 189.

☐ **I asked for help from someone else to complete the review of all or most lessons.**

- You should review the examples and concepts in your Study Notebook and Chapter 4 Foldable.

- Then complete the Chapter 4 Study Guide and Review on pages 186–188 of your textbook.

- If you are unsure of any concepts or skills, refer back to the specific lesson(s).

- You may also want to take the Chapter 4 Practice Test on page 189.

Student Signature

Parent/Guardian Signature

Teacher Signature

# Fractions, Decimals, and Percents

**FOLDABLES** Use the instructions below to make a Foldable to help you organize your notes as you study the chapter. You will see Foldable reminders in the margin of this Interactive Study Notebook to help you in taking notes.

Begin with a sheet of $8\frac{1}{2}''$ by $11''$ construction paper and two sheets of notebook paper.

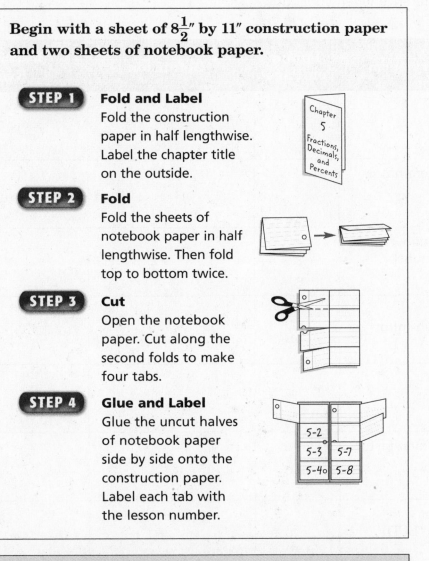

**STEP 1** **Fold and Label**
Fold the construction paper in half lengthwise. Label the chapter title on the outside.

**STEP 2** **Fold**
Fold the sheets of notebook paper in half lengthwise. Then fold top to bottom twice.

**STEP 3** **Cut**
Open the notebook paper. Cut along the second folds to make four tabs.

**STEP 4** **Glue and Label**
Glue the uncut halves of notebook paper side by side onto the construction paper. Label each tab with the lesson number.

**NOTE-TAKING TIP:** Before each lesson, skim through the lesson and write any questions that come to mind in your notes. As you work through the lesson, record the answer to your question.

CHAPTER

5

**BUILD YOUR VOCABULARY**

This is an alphabetical list of new vocabulary terms you will learn in Chapter 5. As you complete the study notes for the chapter, you will see Build Your Vocabulary reminders to complete each term's definition or description on these pages. Remember to add the textbook page number in the second column for reference when you study.

| Vocabulary Term | Found on Page | Definition | Description or Example |
|---|---|---|---|
| bar notation | | | |
| common denominator | | | |
| composite number [kahm-PAH-zuht] | | | |
| equivalent [ih-KWIH-vuh-luhnt] fractions | | | |
| factor tree | | | |
| greatest common factor (GCF) | | | |
| least common denominator (LCD) | | | |
| least common multiple (LCM) | | | |
| multiple | | | |

| Vocabulary Term | Found on Page | Definition | Description or Example |
|---|---|---|---|
| percent | | | |
| prime factorization | | | |
| prime number | | | |
| ratio | | | |
| rational numbers | | | |
| repeating decimals | | | |
| simplest form | | | |
| terminating decimals | | | |
| Venn diagram | | | |

## 5-1 Prime Factorization

**WHAT YOU'LL LEARN**

• Find the prime factorization of a composite number.

---

**BUILD YOUR VOCABULARY** (pages 108–109)

A **prime number** is a whole number greater than 1 that

has exactly [ ] factors, [ ] and [ ].

A **composite number** is a whole number greater than

[ ] that has more than [ ] factors.

Every [ ] number can be written as a product

of prime numbers exactly one way called the **prime factorization**.

A **factor tree** can be used to find the factorization.

---

**FOLDABLES**

## ORGANIZE IT

Under the tab for Lesson 5-1, give examples of prime and composite numbers. Be sure to explain how to tell a prime number from a composite number.

5-2
5-3   5-7
5-4   5-8

---

**EXAMPLES** Identify Numbers as Prime or Composite

**Determine whether each number is *prime* or *composite*.**

**❶ 63**

63 has six factors: 1, [ ], 7, [ ], 21, and [ ].

So, it is [ ].

**❷ 29**

29 has only two factors: [ ] and [ ].

So it is [ ].

**Your Turn** Determine whether each number is *prime* or *composite*.

**a.** 41 [ ]          **b.** 24 [ ]

---

**EXAMPLE** Find the Prime Factorization

**REMEMBER IT**

Multiplication is commutative, so the order of factors does not matter.

**3** Find the prime factorization of 100.

To find the prime factorization, you can use a factor tree or divide by prime numbers. Let's use a factor tree.

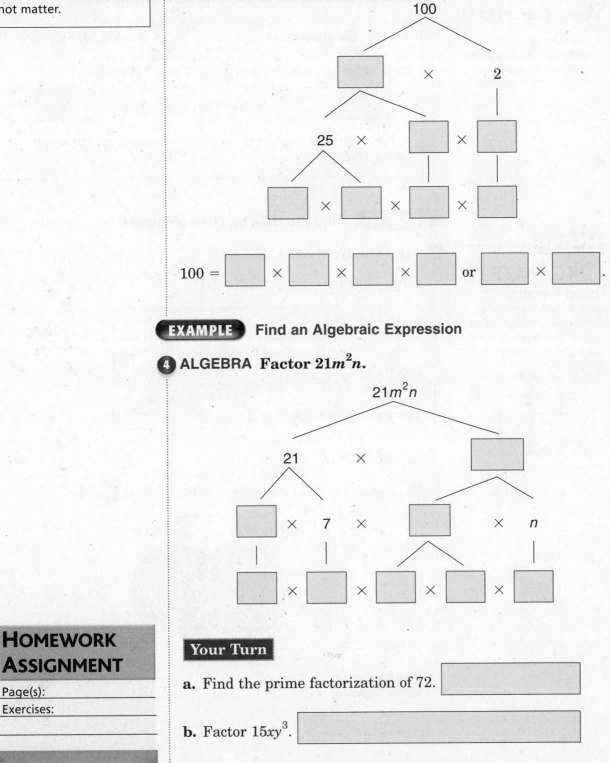

$100 = \boxed{\phantom{0}} \times \boxed{\phantom{0}} \times \boxed{\phantom{0}} \times \boxed{\phantom{0}}$ or $\boxed{\phantom{0}} \times \boxed{\phantom{0}}$.

**EXAMPLE** Find an Algebraic Expression

**4** ALGEBRA Factor $21m^2n$.

**HOMEWORK ASSIGNMENT**

Page(s): _____

Exercises: _____

**Your Turn**

**a.** Find the prime factorization of 72.

**b.** Factor $15xy^3$.

# Greatest Common Factor

**BUILD YOUR VOCABULARY** (pages 108–109)

A **Venn diagram** uses [        ] to show how elements among sets of numbers or objects are related.

The [        ] number that is a common [        ] to two or more numbers is called the **greatest common factor (GCF)**.

**FOLDABLES**

## ORGANIZE IT

Under the tab for Lesson 5-2, take notes on finding the greatest common factor of two or more numbers.

| 5-2 | |
|---|---|
| 5-3 | 5-7 |
| 5-4 | 5-8 |

**EXAMPLE** Find the GCF by Listing Factors

**①  Find the GCF of 28 and 42.**

First, list the factors of 28 and 42.

factors of 28: [        ]

factors of 42: [        ]

The common factors are [        ].

So, the GCF is [    ].

You can draw a Venn diagram to check your answer.

Factors of 28     Factors of 42

4     1  3   6
      2
28    7      21
      14  42

**Your Turn**  Find the GCF of 18 and 45.

[                    ]

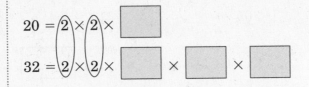

**EXAMPLES** Find GCF Using Prime Factors

**Find the GCF of each set of numbers.**

**2** **20, 32**

**Method 1**   Write the prime factorization.

$20 = \boxed{2} \times \boxed{2} \times \phantom{\square}$

$32 = \boxed{2} \times \boxed{2} \times \phantom{\square} \times \phantom{\square} \times \phantom{\square}$

**Method 2**   Divide by prime numbers.

Divide both 20 and 32 by 2. Then divide the quotients by 2.

$$\begin{array}{r} 5 \quad 8 \\ 2)\overline{10 \quad 16} \end{array}$$

$$2)\overline{10 \quad 32} \quad \longleftarrow \boxed{\text{Start here.}}$$

The common prime factors are 2 and 2.

The GCF of 20 and 32 is $\boxed{\phantom{xx}} \times \boxed{\phantom{xx}}$ or $\boxed{\phantom{xx}}$.

**3** **21, 42, 63**

$21 = \phantom{xx} 3 \times 7$

$42 = 2 \times 3 \times 7$    Circle the common factors.

$63 = 3 \times 3 \times 7$

The common prime factors are 3 and 7.

The GCF is $\boxed{\phantom{xx}} \times \boxed{\phantom{xx}}$, or $\boxed{\phantom{xx}}$.

**Your Turn**   **Find the GCF of each set of numbers.**

**a.** 24, 48, and 60

**b.** 24, 36

---

**EXAMPLE** Find the GCF of an Algebraic Expression

④ **ALGEBRA** Find the GCF of $12p^2$ and $30p^3$.

Factor each expression.

$12p^2 = 2 \cdot$ ②ⓐ · ③ · $p$ · $p$

$30p^3 =$ ② · ③ · 5 · $p$ · $p$ · $p$

The GCF is $2 \cdot 3 \cdot p \cdot p$, or ☐.

**Your Turn** Find the GCF of $14m^2n$ and $21mn^3$.

[blank box]

**EXAMPLE** Use the GCF to Solve a Problem

⑤ **ART** Searra wants to cut a 15-centimeter by 25-centimeter piece of tag board into squares for an art project. She does not want to waste any of the tag board and she wants the largest squares possible. What is the length of the side of the squares she should use?

The largest length of side possible is the GCF of the dimensions of the tag board.

15 = ☐ × ☐

25 = ☐ × ☐

The ☐ of 15 and 25 is ☐. So, Searra should use

squares with sides measuring ☐.

**Your Turn** Alice is making candy baskets using chocolate hearts and lollipops. She has 32 chocolate hearts and 48 lollipops. She wants to have an equal number of chocolate hearts and lollipops in each basket. Find the greatest number of chocolate hearts and lollipops Alice can put in each basket.

[blank box]

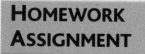

**HOMEWORK ASSIGNMENT**

Page(s):

Exercises:

# Simplifying Fractions

## WHAT YOU'LL LEARN

- Write fractions in simplest form.

**BUILD YOUR VOCABULARY** (pages 108–109)

Fractions having the same [ ] are call **equivalent fractions**.

A fraction is in **simplest form** when the greatest common factor of the [ ] and the [ ] is 1.

**EXAMPLES** Write Fractions in Simplest Form

**Write each fraction in simplest form.**

**1** $\frac{12}{45}$

First, find the GCF of the [ ] and [ ].

factors of 12: [ ]

factors of 45: [ ]

The GCF of 12 and 45 is [ ].

Then, divide the numerator and the denominator by the [ ].

$$\frac{12}{45} = \frac{12 \div \boxed{\phantom{0}}}{45 \div \boxed{\phantom{0}}} = \boxed{\phantom{0}}$$

So, $\frac{12}{45}$ written in simplest form is $\frac{4}{15}$.

### FOLDABLES

## ORGANIZE IT

Under the tab for Lesson 5-3, take notes about simplifying fractions. Be sure to include an example.

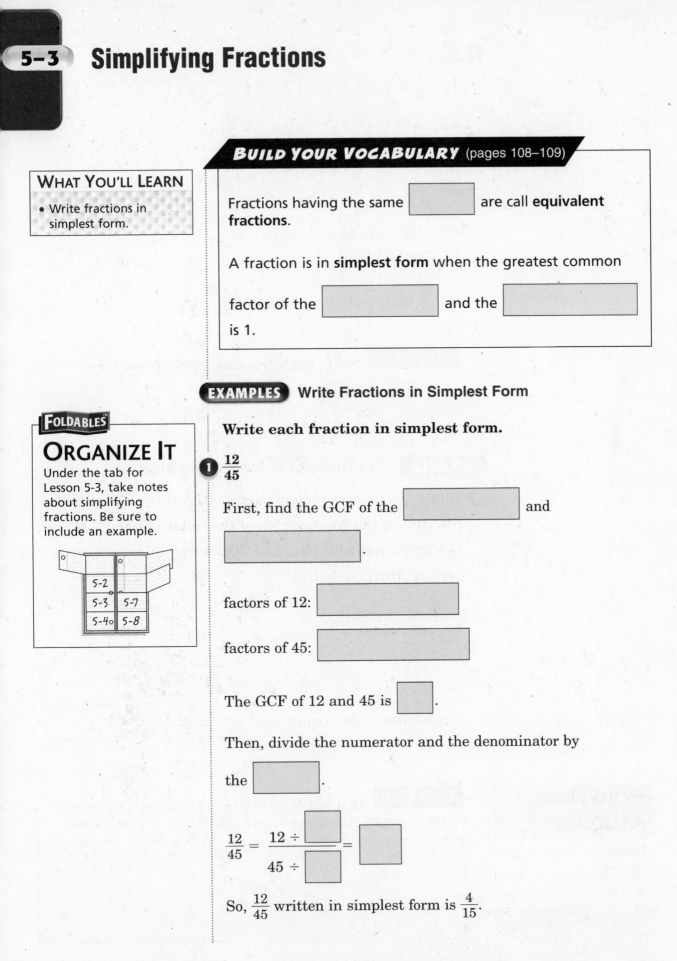

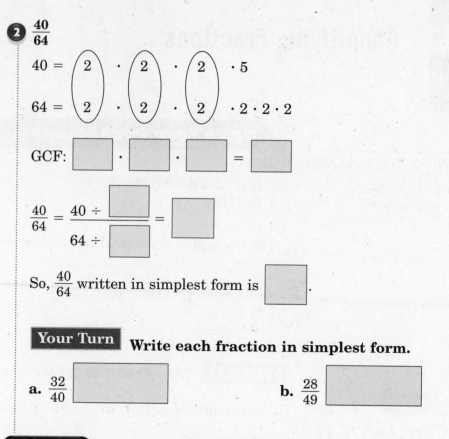

**2** $\frac{40}{64}$

$40 = \boxed{2} \cdot \boxed{2} \cdot \boxed{2} \cdot 5$

$64 = \boxed{2} \cdot \boxed{2} \cdot \boxed{2} \cdot 2 \cdot 2 \cdot 2$

GCF: $\boxed{\phantom{2}} \cdot \boxed{\phantom{2}} \cdot \boxed{\phantom{2}} = \boxed{\phantom{2}}$

$\frac{40}{64} = \dfrac{40 \div \boxed{\phantom{2}}}{64 \div \boxed{\phantom{2}}} = \boxed{\phantom{2}}$

So, $\frac{40}{64}$ written in simplest form is $\boxed{\phantom{2}}$.

**Your Turn** Write each fraction in simplest form.

**a.** $\frac{32}{40}$ $\boxed{\phantom{xxxxxx}}$

**b.** $\frac{28}{49}$ $\boxed{\phantom{xxxxxx}}$

**EXAMPLE** Use Fractions to Solve a Problem

**3** MUSIC Two notes form a *perfect fifth* if the simplified fraction of the frequencies of the notes equals $\frac{3}{4}$. If note D = 294 Hertz and note G = 392 Hertz, do they form a *perfect fifth*?

$\dfrac{\text{frequency of note D}}{\text{frequency of note G}} = \boxed{\phantom{xxx}}$

$= \dfrac{\overset{1}{\cancel{2}} \times 3 \times \overset{1}{\cancel{7}} \times \overset{1}{\cancel{7}}}{\underset{1}{\cancel{2}} \times 2 \times 2 \times \underset{1}{\cancel{7}} \times \underset{1}{\cancel{7}}} = \boxed{\phantom{xxx}}$

The fraction of the frequency of the notes D and G is $\boxed{\phantom{xxx}}$.
So, the two notes do form a *perfect fifth*.

**HOMEWORK ASSIGNMENT**

Page(s): _____

Exercises: _____

_____

**Your Turn** In a bag of 96 marbles, 18 of the marbles are black. Write the fraction of black marbles in simplest form.

# Fractions and Decimals

**EXAMPLES** Write Fractions as Decimals

## WHAT YOU'LL LEARN

- Write fractions as terminating or repeating decimals and write decimals as fractions.

**Write each fraction or mixed number as a decimal.**

**1** $\frac{1}{8}$

The fraction $\frac{1}{8}$ indicates 1 ▭ 8.

**Method 1**   Use paper and pencil.

$$
\begin{array}{r}
0.\boxed{\phantom{000}} \\
8)\overline{1.000} \\
\underline{8}\phantom{00} \\
20\phantom{0} \\
\underline{16}\phantom{0} \\
40 \\
\underline{40} \\
0
\end{array}
$$

Division ends when the denominator is 0.

**Method 2**   Use a ▭.

1 ÷ 8 [ENTER] ▭

So, $\frac{1}{8}$ = ▭

## FOLDABLES

## ORGANIZE IT

Under the tab for Lesson 5-4, take notes on writing fractions as decimals and writing decimals as fractions. Include examples.

5-2
5-3   5-7
5-4   5-8

**2** $7\frac{3}{5}$

**Method 1**   Use paper and pencil.

$7\frac{3}{5} = 7 + \frac{3}{5}$          Write as a sum.

$= 7 + \boxed{\phantom{000}}$          Write $\frac{3}{5}$ as ▭.

$= \boxed{\phantom{000}}$          Add.

**Method 2**   Use a calculator.

3 ÷ 5 + 7 [ENTER] ▭

So, $7\frac{3}{5}$ = ▭

**Your Turn** Write each fraction or mixed number as a decimal.

a. $\frac{2}{5}$ [        ]        b. $3\frac{5}{8}$ [        ]

## BUILD YOUR VOCABULARY (pages 108–109)

A **terminating decimal** is one where the division of the numerator by the denominator has a remainder of [        ].

**Repeating decimals** have a pattern in the digits that repeats [        ].

**Bar notation** is used to indicate that a number repeats forever by writing a [        ] over the [        ] that repeat.

## WRITE IT

Write the following decimal equivalents: $\frac{1}{2}$, $\frac{1}{3}$, $\frac{2}{3}$, $\frac{1}{4}$, $\frac{3}{4}$, $\frac{1}{5}$, $\frac{1}{10}$, $\frac{1}{8}$.

_____

_____

_____

_____

_____

**EXAMPLES** Write Fractions as Repeating Decimals

**3** Write $\frac{1}{11}$ as a decimal.

**Method 1** Use paper and pencil.

$$
\begin{array}{r}
0.0909\ldots \\
11{\overline{)1.0000}} \\
\underline{0}\phantom{.0000} \\
100 \\
[\ \ ] \\
\overline{\phantom{0}10} \\
\underline{0} \\
[\ \ ] \\
\overline{99} \\
[\ \ ]
\end{array}
$$

**Method 2** Use a calculator.

1 [÷] 11 [ENTER] 0.0909 . . .

So, $\frac{1}{11}$ = [        ].

4 Write $6\frac{4}{9}$ as a decimal.

**Method 1**   Use paper and pencil.

$6\frac{4}{9} = 6 + \frac{4}{9}$          Write as a sum.

$= 6 +$ 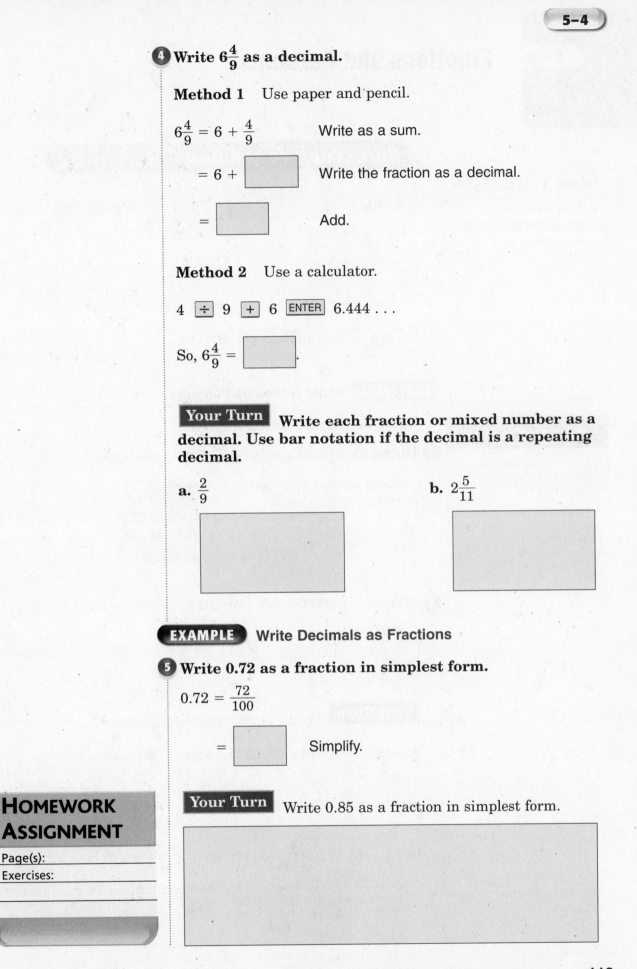          Write the fraction as a decimal.

$=$ [ ]          Add.

**Method 2**   Use a calculator.

4 [÷] 9 [+] 6 [ENTER] 6.444 . . .

So, $6\frac{4}{9} =$ [ ].

[ Your Turn ] **Write each fraction or mixed number as a decimal. Use bar notation if the decimal is a repeating decimal.**

a. $\frac{2}{9}$                                b. $2\frac{5}{11}$

**EXAMPLE**   Write Decimals as Fractions

5 Write 0.72 as a fraction in simplest form.

$0.72 = \frac{72}{100}$

$=$ [ ]          Simplify.

**HOMEWORK ASSIGNMENT**

Page(s):

Exercises:

[ Your Turn ] Write 0.85 as a fraction in simplest form.

## 5–5 Fractions and Percents

**WHAT YOU'LL LEARN**

• Write fractions as percents and percents as fractions.

**BUILD YOUR VOCABULARY** (pages 108–109)

A **ratio** is a [     ] of two numbers by

[     ].

When a [     ] compares a number to [     ],

it can be written as a **percent**.

**EXAMPLES** Write Ratios as Percents

Write each ratio as a percent.

**KEY CONCEPT**

**Percent** A percent is a ratio that compares a number to 100.

**1** Diana scored 63 goals out of 100 attempts.

You can represent 63 out of 100 with a model.

$\frac{63}{100} =$ [     ]

**2** 31.9 out of 100 people bought crunchy peanut butter.

$\frac{31.9}{\boxed{\phantom{00}}} =$ [     ]

**Your Turn** Write each ratio as a percent.

**a.** Alicia sold 34 of the 100 cookies at the bake sale.

**b.** 73.4 out of 100 people preferred the chicken instead of the roast beef.

**120** *Mathematics: Applications and Concepts, Course 2*

**EXAMPLE** Write a Fraction as a Percent

**3** Write $\frac{16}{25}$ as a percent.

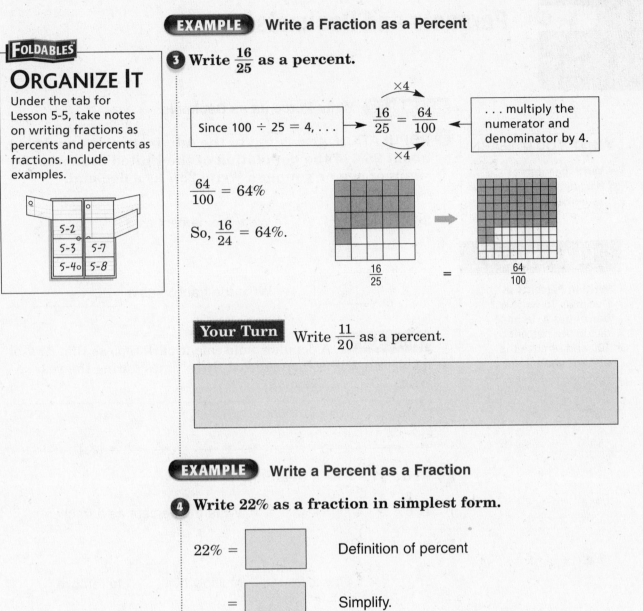

Since $100 \div 25 = 4, \ldots$ → $\frac{16}{25} = \frac{64}{100}$ ← $\ldots$ multiply the numerator and denominator by 4.

$\frac{64}{100} = 64\%$

So, $\frac{16}{24} = 64\%$.

$\frac{16}{25} = \frac{64}{100}$

**Your Turn** Write $\frac{11}{20}$ as a percent.

**EXAMPLE** Write a Percent as a Fraction

**4** Write 22% as a fraction in simplest form.

$22\% = $ ☐ Definition of percent

$= $ ☐ Simplify.

**Your Turn** Write 84% as a fraction in simplest form.

# Percents and Decimals

## WHAT YOU'LL LEARN

- Write percents as decimals and decimals as percents.

## KEY CONCEPT

**Writing Percents as Decimals** To write a percent as a decimal, divide the percent by 100 and remove the percent symbol.

**EXAMPLES** Write Percents as Decimals

① **POPULATION** According to the Administration on Aging, about 28% of the population of the United States is 19 years of age or younger. Write 28% as a decimal.

$$28\% = \frac{28}{\boxed{\phantom{0}}}$$   Write the percent as a $\boxed{\phantom{00}}$.

$$= \boxed{\phantom{0}}$$   Write the fraction as a $\boxed{\phantom{00}}$.

**Your Turn** A popular amusement park reports that 17% of its visitors will return at least three times during the year. Write 17% as a decimal.

② Write 47.8% as a decimal.

$$47.8\% = \frac{47.8}{\boxed{\phantom{0}}}$$   Write the percent as a fraction.

$$= \frac{47.8 \times \boxed{\phantom{0}}}{\boxed{\phantom{0}} \times \boxed{\phantom{0}}}$$   Multiply by $\boxed{\phantom{0}}$ to remove the decimal in the numerator.

$$= \frac{\boxed{\phantom{0}}}{1,000}$$   Simplify.

$$= \boxed{\phantom{0}}$$   Write the fraction as a decimal.

So, $47.8\% = \boxed{\phantom{00}}$.

**Your Turn** Write 83.2% as a decimal.

**EXAMPLES** Write Percents as Decimals

**3** Write 95.3% as a decimal.

95.3% = 95.3      Divide by [ ].

= [ ]      Remove the [ ].

So, 95.3% = [ ].

**4** Write $8\frac{1}{5}\%$ as a decimal.

$8\frac{1}{5}\%$ = [ ]      Write $\frac{1}{5}$ as [ ].

= 008.2      [ ] by 100.

= [ ]      Remove the [ ].

So, $8\frac{1}{5}\%$ = [ ].

**Your Turn** Write each percent as a decimal.

**a.** 38% [ ]      **b.** $27\frac{3}{4}\%$ [ ]

**EXAMPLES** Write Decimals as Percents

**5** POPULATION In 1790, about 0.05 of the population of the United States lived in an urban setting. Write 0.05 as a percent.

0.05 = [ ]      Definition of decimal

= [ ]      Definition of [ ]

**Your Turn** In 2000, the population of Illinois had increased by 0.086 from 1990. Write 0.086 as a percent.

[ ]

**6** **Write 0.121 as a percent.**

$$0.121 = \frac{121}{\boxed{\phantom{xxx}}}$$   Definition of decimal

$$= \boxed{\phantom{xxxxx}}$$   Divide both numerator and denominator by 10.

$$= \boxed{\phantom{xxxxx}}$$   Definition of percent

So, $0.121 = \boxed{\phantom{xxxxx}}$ .

**Your Turn**   Write 0.364 as a percent.

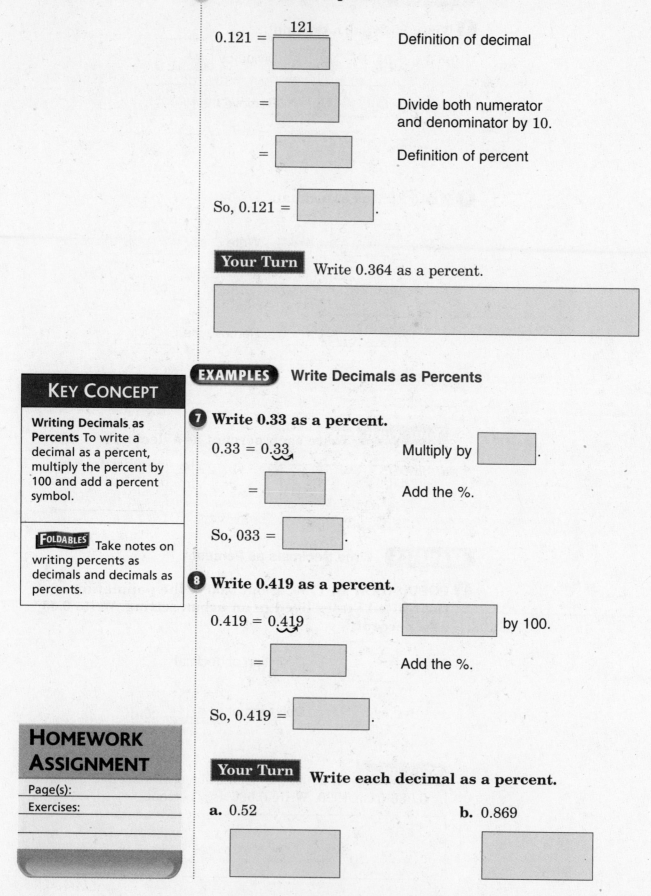

## KEY CONCEPT

**Writing Decimals as Percents** To write a decimal as a percent, multiply the percent by 100 and add a percent symbol.

**FOLDABLES** Take notes on writing percents as decimals and decimals as percents.

**EXAMPLES**   **Write Decimals as Percents**

**7** **Write 0.33 as a percent.**

$$0.33 = 0.33\overset{\frown}{\phantom{x}}$$   Multiply by $\boxed{\phantom{xx}}$ .

$$= \boxed{\phantom{xxxx}}$$   Add the %.

So, $033 = \boxed{\phantom{xxx}}$ .

**8** **Write 0.419 as a percent.**

$$0.419 = 0.419\overset{\frown}{\phantom{x}}$$   $\boxed{\phantom{xxxxx}}$ by 100.

$$= \boxed{\phantom{xxxx}}$$   Add the %.

So, $0.419 = \boxed{\phantom{xxx}}$ .

## HOMEWORK ASSIGNMENT

Page(s):

Exercises:

**Your Turn**   Write each decimal as a percent.

**a.** 0.52

**b.** 0.869

**Least Common Multiple**

## WHAT YOU'LL LEARN

- Find the least common multiple of two or more numbers.

**BUILD YOUR VOCABULARY** (pages 108–109)

A **multiple** is the [ ] of a number and any

[ ] number.

The **least common denominator (LCM)** of two or more

numbers is the [ ] of their common multiples,

excluding [ ].

**EXAMPLE** Find the LCM by Listing Multiples

**FOLDABLES**

## ORGANIZE IT

Under the tab for Lesson 5-7, take notes about least common multiples. Be sure to include examples.

5-2
5-3  5-7
5-4  5-8

① **Find the LCM of 4 and 6.**

First, list the multiples of 4 and 6.

multiples of 4:

[ ]

multiples of 6:

[ ]

The common multiples are [ ], 24, 36 . . . .

The LCM of 4 and 6 is [ ].

**Your Turn** Find the LCM of 8 and 12.

[ ]

**EXAMPLES** Find the LCM Using Prime Factors

**2** Find the LCM of 4 and 15.

Write the prime factorization.

4 = [　] × 2 or [　]

15 = [　] × [　]

The prime factors of 4 and 15 are [　].

The LCM of 4 and 15 is [　] × 3 × 5, or [　].

**3** Find the LCM of 18, 24, and 48.

$18 = 2 \times \boxed{3^2}$

$24 = 23 \times 3$

$48 = \boxed{2^4} \times 3$

LCM: [　] × [　] = [　]

The LCM of 18, 24, and 48 is [　].

**Your Turn** Find the LCM of each set of numbers.

**a.** 6, 14

**b.** 12, 20, 45

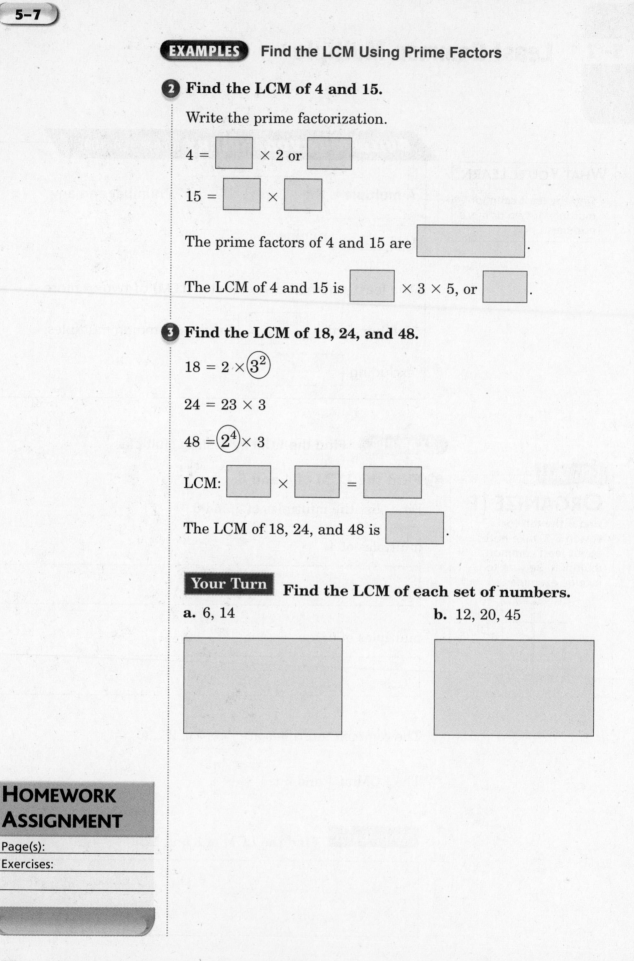

**HOMEWORK ASSIGNMENT**

Page(s): _____

Exercises: _____

_____

# Comparing and Ordering Rational Numbers

**WHAT YOU'LL LEARN**

- Compare and order fractions, decimals, and percents.

**BUILD YOUR VOCABULARY** (pages 108–109)

A **common denominator** is a common multiple of two or more _____ .

The **least common denominator (LCD)** is the [    ] of the denominators.

**Rational numbers** are numbers that can be written as fractions and include fractions, terminating and repeating decimals, and integers.

**REVIEW IT**

Explain how to write $\frac{48}{60}$ as a decimal.
(*Lesson 5-4*)

_____

_____

_____

_____

**EXAMPLE** Compare Fractions

❶ **GRADES** Enrique and his younger brother both had a math test last Friday. Enrique scored 48 points out of 60 and his brother scored 55 points out of 75. Who got the better score, Enrique or his brother?

**Method 1** Rename using the LCD. Then compare.

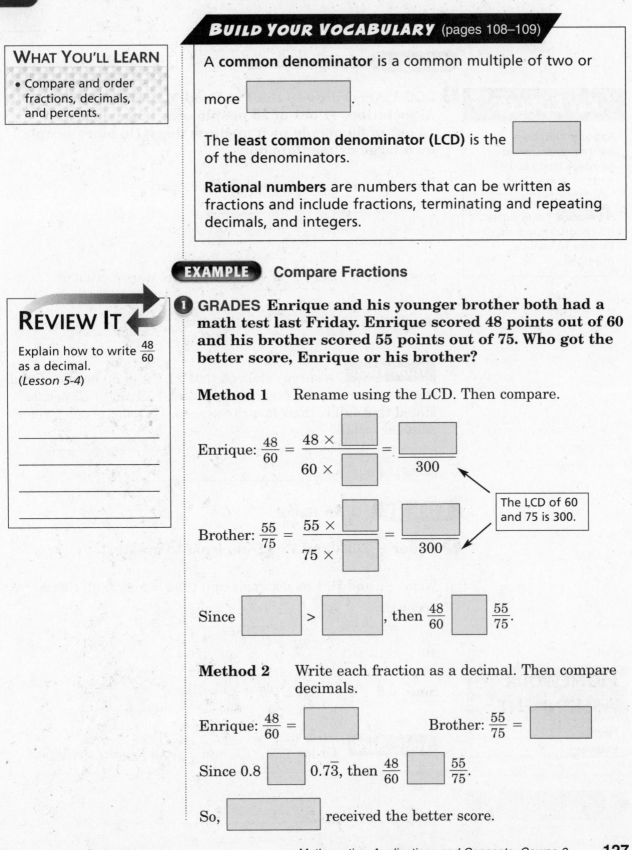

Enrique: $\dfrac{48}{60} = \dfrac{48 \times \boxed{\phantom{0}}}{60 \times \boxed{\phantom{0}}} = \dfrac{\boxed{\phantom{0}}}{300}$

Brother: $\dfrac{55}{75} = \dfrac{55 \times \boxed{\phantom{0}}}{75 \times \boxed{\phantom{0}}} = \dfrac{\boxed{\phantom{0}}}{300}$

The LCD of 60 and 75 is 300.

Since $\boxed{\phantom{0}} > \boxed{\phantom{0}}$ , then $\dfrac{48}{60} \boxed{\phantom{0}} \dfrac{55}{75}$.

**Method 2** Write each fraction as a decimal. Then compare decimals.

Enrique: $\dfrac{48}{60} = \boxed{\phantom{0}}$          Brother: $\dfrac{55}{75} = \boxed{\phantom{0}}$

Since $0.8 \boxed{\phantom{0}} 0.7\overline{3}$, then $\dfrac{48}{60} \boxed{\phantom{0}} \dfrac{55}{75}$.

So, $\boxed{\phantom{0}}$ received the better score.

**Your Turn** During the hockey season, Kyle scored 14 goals out of 24 shots taken and his teammate David scored 18 goals out of 30 shots taken. Who had the higher scoring percentage?

**EXAMPLE** Compare Ratios

**2** DOGS According to the Pet Food Manufacturer's Association, 11 out of 25 people own large dogs and 13 out of 50 people own medium dogs. Do more people own large or medium dogs?

Write $\frac{11}{25}$ and $\frac{13}{50}$ as decimals and compare.

$\frac{11}{25} = $ ☐      $\frac{13}{50} = $ ☐

Since $0.44 > 0.26$, $\frac{11}{25}$ ☐ $\frac{13}{50}$. So, a greater fraction of

people own ☐ dogs than own ☐ dogs.

**Your Turn** A survey showed that 21 out of 50 people stated that summer is their favorite season and 13 out of 25 people stated that fall is their favorite season. Do more people prefer summer or fall?

**EXAMPLE** Order Ratios

**3** Order $\frac{7}{10}$, 0.6, and 72% from least to greatest.

Write $\frac{7}{10}$ and 72% as decimals and then compare all three decimals.

$\frac{7}{10} = $ ☐      $72\% = $ ☐

Since $0.6 < 0.7 < 0.72$, $0.6 < $ ☐ $ < $ ☐.

**HOMEWORK ASSIGNMENT**

Page(s):
Exercises:

**Your Turn** Order 37%, 0.3, and $\frac{3}{8}$ from least to greatest.

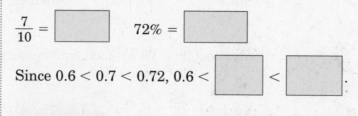

# BRINGING IT ALL TOGETHER

## STUDY GUIDE

| **FOLDABLES™** | VOCABULARY PUZZLEMAKER | BUILD YOUR VOCABULARY |
|---|---|---|
| Use your **Chapter 5 Foldable** to help you study for your chapter test. | To make a crossword puzzle, word search, or jumble puzzle of the vocabulary words in Chapter 5, go to: www.glencoe.com/sec/math/ t_resources/free/index.php | You can use your completed **Vocabulary Builder** (*pages 108–109*) to help you solve the puzzle. |

### 5-1

### Prime Factorization

**Underline the correct terms to complete each sentence.**

1. A factor tree is complete when all of the factors at the bottom of the factor tree are (*prime, composite*) factors.

2. The order of the factors in prime factorization (*does, does not*) matter.

**Find the prime factorization of each number.**

3. 36

4. 48

5. 250

6. 60

### 5-2

### Greatest Common Factor

**Complete each sentence.**

7. A _____ shows how elements of sets of numbers are related.

8. A prime factor is a factor that is a _____ number.

**9.** You can find the [____] of two numbers by

[_____] the common prime factors.

**Find the common prime factors and GCF of each set of numbers.**

**10.** 20, 24 [_____]          **11.** 28, 42 [_____]

**5-3**

## Simplifying Fractions

**Complete each sentence.**

**12.** To find the simplest form of a fraction, [____] the

numerator and the denominator by the [____].

**13.** To check if a fraction is in simplest form, [____] the

numerator and the [_____] by the [____], and

if you get your original [_____] the answer is correct.

**Write each fraction in simplest form.**

**14.** $\frac{18}{24}$ [__]          **15.** $\frac{15}{60}$ [__]

**5-4**

## Fractions and Decimals

**Write each fraction or mixed number as a decimal. Use bar notation if the decimal is a repeating decimal.**

**16.** $3\frac{2}{3}$ [__]     **17.** $5\frac{3}{4}$ [__]     **18.** $\frac{2}{5}$ [__]

**19.** $7\frac{3}{8}$ [__]     **20.** $6\frac{1}{2}$ [__]     **21.** $\frac{7}{10}$ [__]

## 5-5
## Fractions and Percents

**22.** Write the ratio that compares 4 to 25 in three different ways.

**23.** Write the ratio in exercise 23 as a percent.

**24.** Write 88% as a fraction in simplest form.

**25.** Write $\frac{9}{20}$ as a percent.

## 5-6
## Percents and Decimals

**Write each percent as a decimal.**

**26.** 69%       **27.** 3%       **28.** $32\frac{1}{4}\%$

**Write each decimal as a percent.**

**29.** 0.47       **30.** 0.5775       **31.** 0.09

## 5-7
## Least Common Multiple

**Find the LCM of each set of numbers.**

**32.** 15, 36       **33.** 21, 70

**34.** 16, 20       **35.** 6, 9, 24

**36.** 12, 18, 30       **37.** 14, 28, 35

## 5-8
## Comparing and Ordering Rational Numbers

**Replace each ● with <, >, or = to make each sentence true.**

**38.** $\frac{14}{35}$ ● $\frac{12}{20}$       **39.** $\frac{21}{49}$ ● $\frac{18}{63}$

# CHAPTER 5

## Checklist

# ARE YOU READY FOR THE CHAPTER TEST?

# Math Online

Visit **msmath2.net** to access your textbook, more examples, self-check quizzes, and practice tests to help you study the concepts in Chapter 5.

**Check the one that applies. Suggestions to help you study are given with each item.**

**☐ I completed the review of all or most lessons without using my notes or asking for help.**

- You are probably ready for the Chapter Test.

- You may want to take the Chapter 5 Practice Test on page 235 of your textbook as a final check.

**☐ I used my Foldable or Study Notebook to complete the review of all or most lessons.**

- You should complete the Chapter 5 Study Guide and Review on pages 232–234 of your textbook.

- If you are unsure of any concepts or skills, refer back to the specific lesson(s).

- You may also want to take the Chapter 5 Practice Test on page 235 of your textbook.

**☐ I asked for help from someone else to complete the review of all or most lessons.**

- You should review the examples and concepts in your Study Notebook and Chapter 5 Foldable.

- Then complete the Chapter 5 Study Guide and Review on pages 232–234 of your textbook.

- If you are unsure of any concepts or skills, refer back to the specific lesson(s).

- You may also want to take the Chapter 5 Practice Test on page 235 of your textbook.

Student Signature          Parent/Guardian Signature

Teacher Signature

# Applying Fractions

**FOLDABLES** Use the instructions below to make a Foldable to help you organize your notes as you study the chapter. You will see Foldable reminders in the margin of this Interactive Study Notebook to help you in taking notes.

Begin with a sheet of $8\frac{1}{2}$" by 11" paper, four index cards, and glue.

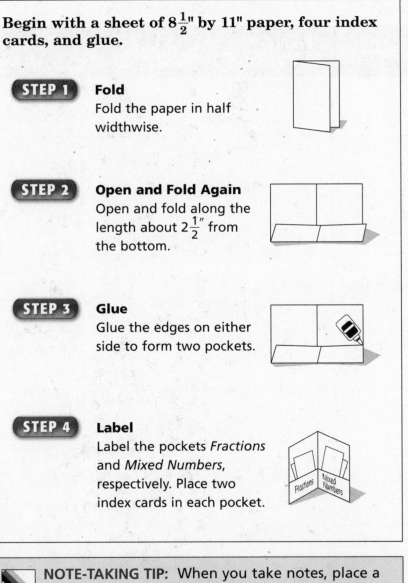

**STEP 1**  **Fold**
Fold the paper in half widthwise.

**STEP 2**  **Open and Fold Again**
Open and fold along the length about $2\frac{1}{2}$" from the bottom.

**STEP 3**  **Glue**
Glue the edges on either side to form two pockets.

**STEP 4**  **Label**
Label the pockets *Fractions* and *Mixed Numbers*, respectively. Place two index cards in each pocket.

**NOTE-TAKING TIP:** When you take notes, place a question mark next to any concepts you do not understand. Be sure to ask your teacher to clarify these concepts before a test.

Chapter 6

## BUILD YOUR VOCABULARY

This is an alphabetical list of new vocabulary terms you will learn in Chapter 6. As you complete the study notes for the chapter, you will see Build Your Vocabulary reminders to complete each term's definition or description on these pages. Remember to add the textbook page number in the second column for reference when you study.

| Vocabulary Term | Found on Page | Definition | Description or Example |
|---|---|---|---|
| area | | | |
| center | | | |
| circle | | | |
| circumference [suhr-KUHM-fuh-ruhns] | | | |
| compatible numbers | | | |
| cup | | | |
| diameter [deye-A-muh-tuhr] | | | |
| formula [FOHR-myuh-luh] | | | |
| gallon | | | |

| Vocabulary Term | Found on Page | Definition | Description or Example |
|---|---|---|---|
| multiplicative inverse [MUHL-tuh-PLIH-kuh-tihv] | | | |
| ounce | | | |
| perimeter [puh-RIH-muh-tuhr] | | | |
| pint | | | |
| pound | | | |
| quart | | | |
| radius [RAY-dee-uhs] | | | |
| reciprocal [rih-SIH-pruh-kuhl | | | |
| ton [TUHN] | | | |

**Estimating with Fractions**

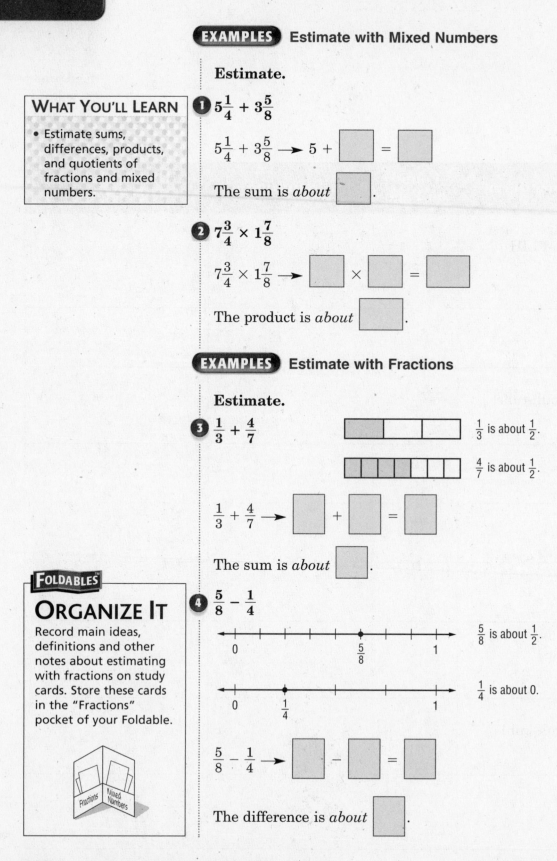

**EXAMPLES** Estimate with Mixed Numbers

**Estimate.**

① $5\frac{1}{4} + 3\frac{5}{8}$

$5\frac{1}{4} + 3\frac{5}{8} \longrightarrow 5 + \boxed{\phantom{0}} = \boxed{\phantom{0}}$

The sum is *about* $\boxed{\phantom{0}}$.

② $7\frac{3}{4} \times 1\frac{7}{8}$

$7\frac{3}{4} \times 1\frac{7}{8} \longrightarrow \boxed{\phantom{0}} \times \boxed{\phantom{0}} = \boxed{\phantom{0}}$

The product is *about* $\boxed{\phantom{0}}$.

**EXAMPLES** Estimate with Fractions

**Estimate.**

③ $\frac{1}{3} + \frac{4}{7}$

$\frac{1}{3}$ is about $\frac{1}{2}$.

$\frac{4}{7}$ is about $\frac{1}{2}$.

$\frac{1}{3} + \frac{4}{7} \longrightarrow \boxed{\phantom{0}} + \boxed{\phantom{0}} = \boxed{\phantom{0}}$

The sum is *about* $\boxed{\phantom{0}}$.

④ $\frac{5}{8} - \frac{1}{4}$

$\frac{5}{8}$ is about $\frac{1}{2}$.

$\frac{1}{4}$ is about 0.

$\frac{5}{8} - \frac{1}{4} \longrightarrow \boxed{\phantom{0}} - \boxed{\phantom{0}} = \boxed{\phantom{0}}$

The difference is *about* $\boxed{\phantom{0}}$.

**WHAT YOU'LL LEARN**

• Estimate sums, differences, products, and quotients of fractions and mixed numbers.

**FOLDABLES**

**ORGANIZE IT**

Record main ideas, definitions and other notes about estimating with fractions on study cards. Store these cards in the "Fractions" pocket of your Foldable.

136    *Mathematics: Applications and Concepts, Course 2*

# REMEMBER IT

Some fractions are easy to round because they are close to 1. Examples of these kinds of fractions are ones where the numerator is one less than the denominator, such as $\frac{4}{5}$ or $\frac{7}{8}$.

**Your Turn** Estimate.

**a.** $2\frac{7}{9} + 5\frac{1}{4}$

**b.** $4\frac{2}{3} \times 3\frac{1}{8}$

**c.** $\frac{8}{9} + \frac{1}{6}$

**d.** $\frac{11}{12} - \frac{2}{9}$

**e.** $\frac{2}{3} + \frac{7}{8}$

## BUILD YOUR VOCABULARY (page 134)

Numbers that are easy to compute [ ] are called **compatible numbers**.

## EXAMPLES Use Compatible Numbers

Estimate.

**5** $\frac{3}{4} \times 21$

$\frac{3}{4} \times 21 \longrightarrow \frac{3}{4} \times 20 =$ [ ]   Round 21 to 20, since 20 is divisible by 4.

The product is *about* [ ].

**6** $15\frac{8}{9} \div 3\frac{2}{5}$

$15\frac{8}{9} \div 3\frac{2}{5} \longrightarrow 16 \div 3\frac{2}{5}$   Round $15\frac{8}{9}$ to 16.

$\longrightarrow 16 \div 4$   Round $3\frac{2}{5}$ to 4, since 16 is divisible by 4.

The quotient is *about* [ ].

# HOMEWORK ASSIGNMENT

Page(s):

Exercises:

**Your Turn** Estimate.

**a.** $\frac{2}{3} \times 17$

**b.** $20\frac{1}{4} \div 3\frac{2}{3}$

# 6-2 Adding and Subtracting Fractions

## WHAT YOU'LL LEARN

• Add and subtract fractions.

## KEY CONCEPT

**Adding and Subtracting Like Fractions** To add or subtract like fractions, add or subtract the numerators and write the result over the denominator. Simplify if necessary.

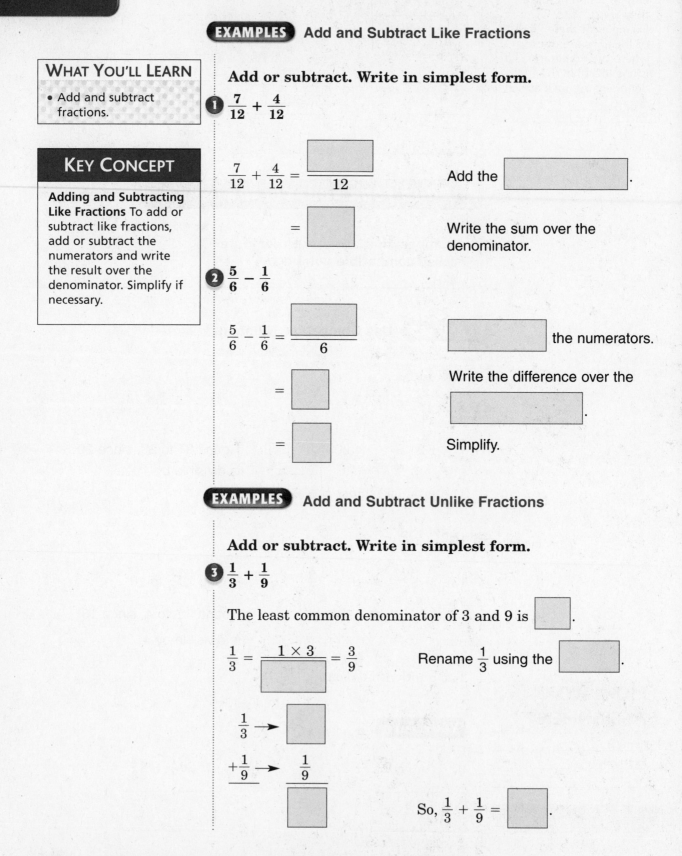

**EXAMPLES** Add and Subtract Like Fractions

Add or subtract. Write in simplest form.

① $\frac{7}{12} + \frac{4}{12}$

$$\frac{7}{12} + \frac{4}{12} = \frac{\boxed{\phantom{xx}}}{12}$$   Add the $\boxed{\phantom{xxxxxx}}$.

$$= \boxed{\phantom{xx}}$$   Write the sum over the denominator.

② $\frac{5}{6} - \frac{1}{6}$

$$\frac{5}{6} - \frac{1}{6} = \frac{\boxed{\phantom{xx}}}{6}$$   $\boxed{\phantom{xxxxxx}}$ the numerators.

$$= \boxed{\phantom{x}}$$   Write the difference over the $\boxed{\phantom{xxxxx}}$.

$$= \boxed{\phantom{x}}$$   Simplify.

**EXAMPLES** Add and Subtract Unlike Fractions

Add or subtract. Write in simplest form.

③ $\frac{1}{3} + \frac{1}{9}$

The least common denominator of 3 and 9 is $\boxed{\phantom{x}}$.

$$\frac{1}{3} = \frac{1 \times 3}{\boxed{\phantom{xx}}} = \frac{3}{9}$$   Rename $\frac{1}{3}$ using the $\boxed{\phantom{xxx}}$.

$$\frac{1}{3} \rightarrow \boxed{\phantom{xx}}$$

$$+\frac{1}{9} \rightarrow \frac{1}{9}$$
_____
$$\boxed{\phantom{xx}}$$   So, $\frac{1}{3} + \frac{1}{9} = \boxed{\phantom{x}}$.

Explain what happens to denominators when adding like fractions.

_____

_____

_____

_____

_____

**4** $\frac{3}{4} - \frac{1}{6}$

The LCD of 4 and 6 is ☐ .

$\frac{3}{4} \rightarrow \frac{3 \times 3}{4 \times 3} \rightarrow \dfrac{\ }{12}$

Rename each fraction using the LCD.

$-\frac{1}{6} \rightarrow \frac{1 \times 2}{6 \times 2} \rightarrow \dfrac{\ }{12}$

So, $\frac{3}{4} - \frac{1}{6} =$ ☐ .

**Your Turn** **Add or subtract. Write in simplest form.**

**a.** $\frac{7}{15} + \frac{4}{15}$

**b.** $\frac{7}{8} - \frac{5}{8}$

**c.** $\frac{3}{8} + \frac{1}{4}$

**d.** $\frac{7}{9} - \frac{1}{6}$

**HOMEWORK ASSIGNMENT**

Page(s):

Exercises:

# Adding and Subtracting Mixed Numbers

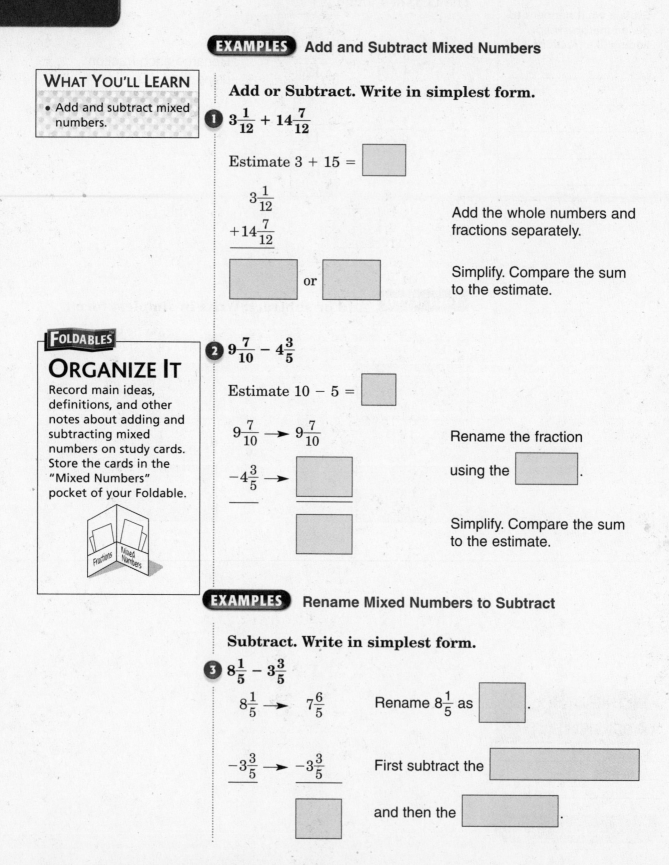

**EXAMPLES** Add and Subtract Mixed Numbers

## WHAT YOU'LL LEARN

- Add and subtract mixed numbers.

**Add or Subtract. Write in simplest form.**

1. $3\frac{1}{12} + 14\frac{7}{12}$

Estimate $3 + 15 = $ ▭

$$3\frac{1}{12}$$
$$+14\frac{7}{12}$$

Add the whole numbers and fractions separately.

▭ or ▭

Simplify. Compare the sum to the estimate.

**FOLDABLES**

## ORGANIZE IT

Record main ideas, definitions, and other notes about adding and subtracting mixed numbers on study cards. Store the cards in the "Mixed Numbers" pocket of your Foldable.

2. $9\frac{7}{10} - 4\frac{3}{5}$

Estimate $10 - 5 = $ ▭

$9\frac{7}{10} \rightarrow 9\frac{7}{10}$

$-4\frac{3}{5} \rightarrow$ ▭

Rename the fraction

using the ▭.

▭

Simplify. Compare the sum to the estimate.

**EXAMPLES** Rename Mixed Numbers to Subtract

**Subtract. Write in simplest form.**

3. $8\frac{1}{5} - 3\frac{3}{5}$

$8\frac{1}{5} \rightarrow 7\frac{6}{5}$

Rename $8\frac{1}{5}$ as ▭.

$-3\frac{3}{5} \rightarrow -3\frac{3}{5}$

First subtract the ▭

▭

and then the ▭.

**4** $11\frac{5}{9} - 8\frac{2}{3}$

First, rename the fractions. Then rename $11\frac{5}{9}$.

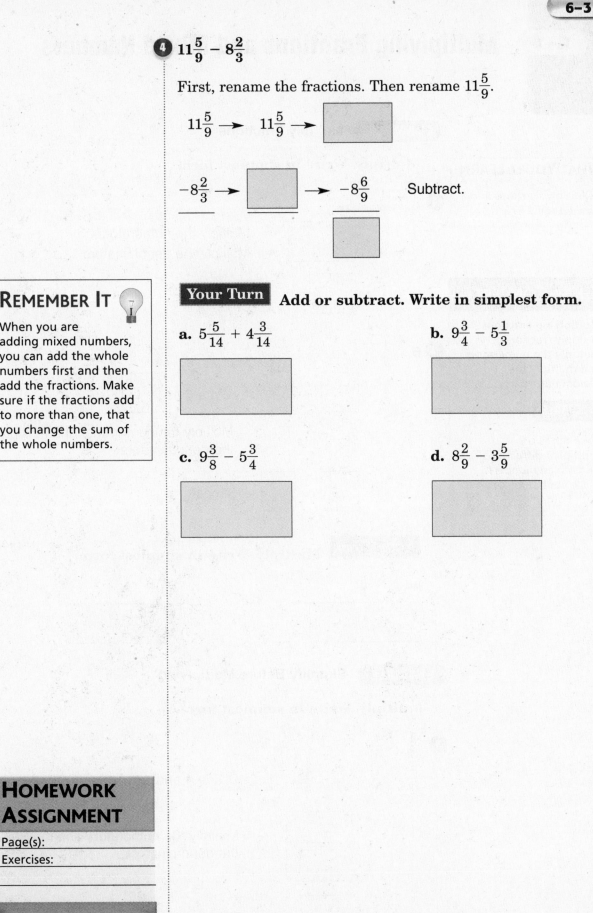

$11\frac{5}{9} \rightarrow 11\frac{5}{9} \rightarrow$ [ ]

$-8\frac{2}{3} \rightarrow$ [ ] $\rightarrow -8\frac{6}{9}$  Subtract.

[ ]

**REMEMBER IT**

When you are adding mixed numbers, you can add the whole numbers first and then add the fractions. Make sure if the fractions add to more than one, that you change the sum of the whole numbers.

**Your Turn**  **Add or subtract. Write in simplest form.**

**a.** $5\frac{5}{14} + 4\frac{3}{14}$

**b.** $9\frac{3}{4} - 5\frac{1}{3}$

**c.** $9\frac{3}{8} - 5\frac{3}{4}$

**d.** $8\frac{2}{9} - 3\frac{5}{9}$

**HOMEWORK ASSIGNMENT**

Page(s):

Exercises:

# Multiplying Fractions and Mixed Numbers

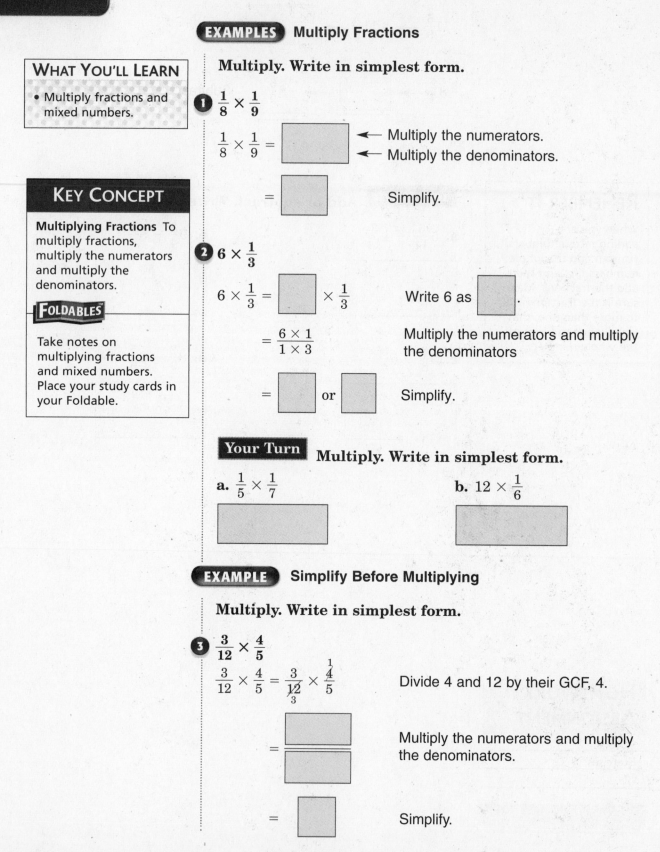

**WHAT YOU'LL LEARN**

- Multiply fractions and mixed numbers.

**KEY CONCEPT**

**Multiplying Fractions** To multiply fractions, multiply the numerators and multiply the denominators.

**FOLDABLES**

Take notes on multiplying fractions and mixed numbers. Place your study cards in your Foldable.

**EXAMPLES** Multiply Fractions

**Multiply. Write in simplest form.**

**1** $\frac{1}{8} \times \frac{1}{9}$

$\frac{1}{8} \times \frac{1}{9} =$ ⬅ Multiply the numerators.
⬅ Multiply the denominators.

$=$ ☐ Simplify.

**2** $6 \times \frac{1}{3}$

$6 \times \frac{1}{3} =$ ☐ $\times \frac{1}{3}$ Write 6 as ☐.

$= \frac{6 \times 1}{1 \times 3}$ Multiply the numerators and multiply the denominators

$=$ ☐ or ☐ Simplify.

**Your Turn** Multiply. Write in simplest form.

**a.** $\frac{1}{5} \times \frac{1}{7}$

**b.** $12 \times \frac{1}{6}$

**EXAMPLE** Simplify Before Multiplying

**Multiply. Write in simplest form.**

**3** $\frac{3}{12} \times \frac{4}{5}$

$\frac{3}{12} \times \frac{4}{5} = \frac{3}{\underset{3}{12}} \times \frac{\overset{1}{4}}{5}$ Divide 4 and 12 by their GCF, 4.

$=$ ☐ Multiply the numerators and multiply the denominators.

$=$ ☐ Simplify.

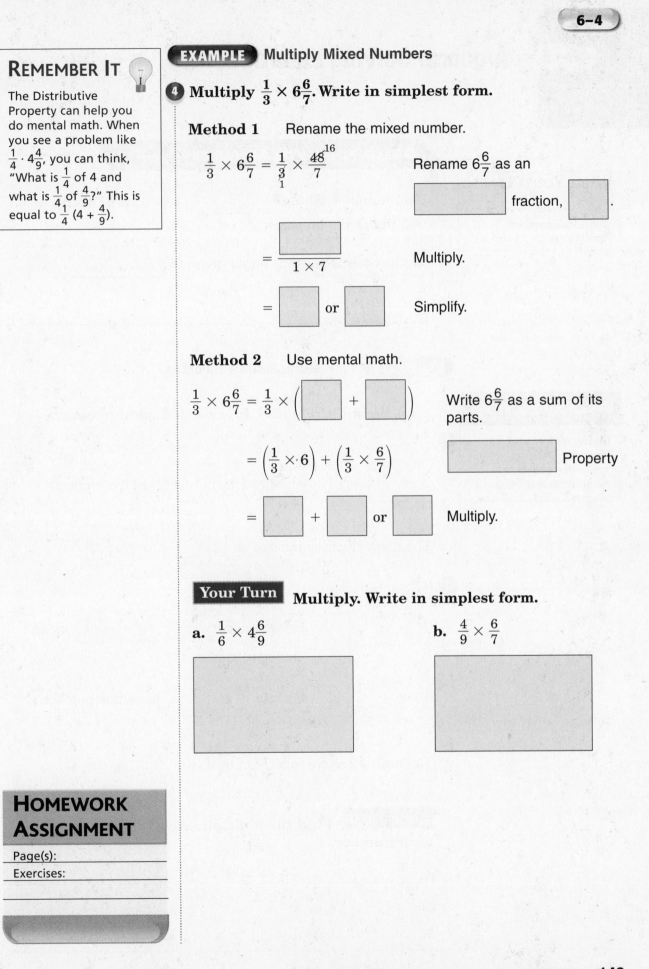

**EXAMPLE** **Multiply Mixed Numbers**

4 **Multiply $\frac{1}{3} \times 6\frac{6}{7}$. Write in simplest form.**

**Method 1**     Rename the mixed number.

$$\frac{1}{3} \times 6\frac{6}{7} = \frac{1}{\underset{1}{3}} \times \frac{\overset{16}{48}}{7}$$     Rename $6\frac{6}{7}$ as an

[    ] fraction, [  ].

$$= \frac{\boxed{\phantom{xx}}}{1 \times 7}$$     Multiply.

$$= \boxed{\phantom{xx}} \text{ or } \boxed{\phantom{xx}}$$     Simplify.

**Method 2**     Use mental math.

$$\frac{1}{3} \times 6\frac{6}{7} = \frac{1}{3} \times \left( \boxed{\phantom{xx}} + \boxed{\phantom{xx}} \right)$$     Write $6\frac{6}{7}$ as a sum of its parts.

$$= \left( \frac{1}{3} \times 6 \right) + \left( \frac{1}{3} \times \frac{6}{7} \right)$$     [         ] Property

$$= \boxed{\phantom{xx}} + \boxed{\phantom{xx}} \text{ or } \boxed{\phantom{xx}}$$     Multiply.

**Your Turn** **Multiply. Write in simplest form.**

a.  $\frac{1}{6} \times 4\frac{6}{9}$

b.  $\frac{4}{9} \times \frac{6}{7}$

**HOMEWORK ASSIGNMENT**

Page(s): _____

Exercises: _____

_____

_____

# Algebra: Solving Equations

**WHAT YOU'LL LEARN**

- Solve equations with rational number solutions.

**BUILD YOUR VOCABULARY** (page 135)

Two numbers whose [ ] is [ ] are called **multiplicative inverses.**

**Reciprocals** is another name given to [ ] [ ].

**EXAMPLES** Find Multiplicative Inverses

**KEY CONCEPT**

**Multiplicative Inverse Property** The product of a number and its multiplicative inverse is 1.

Find the multiplicative inverse of each number.

**1** $\dfrac{4}{7}$

$\dfrac{4}{7} \cdot$ [ ] $= 1$     Multiply $\dfrac{4}{7}$ by [ ] to get the product 1.

The multiplicative inverse of $\dfrac{4}{7}$ is [ ], or [ ].

**2** $6\dfrac{1}{4}$

$6\dfrac{1}{4} =$ [ ]     Rename the [ ] as an improper fraction.

$\dfrac{25}{4} \cdot$ [ ] $= 1$     Multiply $\dfrac{25}{4}$ by [ ] to get the product 1.

The multiplicative inverse of $6\dfrac{1}{4}$ is [ ].

**Your Turn** Find the multiplicative inverse of each number.

a. $\dfrac{5}{8}$

b. $4\dfrac{1}{3}$

**EXAMPLE**   Solve a Division Equation

**3** Solve $\frac{p}{6} = 11$.

$\frac{p}{6} = 11$   Write the equation.

$\frac{p}{6} \cdot \boxed{\phantom{xx}} = 11 \cdot \boxed{\phantom{xx}}$   Multiply each side by $\boxed{\phantom{xx}}$.

$p = \boxed{\phantom{xxx}}$   Simplify.

The solution is $\boxed{\phantom{xxx}}$.

**EXAMPLE**   Use a Reciprocal to Solve an Equation

**4** Solve $\frac{4}{5}y = -8$.

$\frac{4}{5}y = 28$   Write the equation.

$\boxed{\phantom{xx}} \, \frac{4}{5}y = \boxed{\phantom{xx}} (-8)$   Multiply each side by the

$\boxed{\phantom{xxxx}}$ of $\frac{4}{5}$.

$y = \boxed{\phantom{xx}}$ or $\boxed{\phantom{xx}}$   Simplify.

**Your Turn**   Solve.

a. $\frac{m}{9} = 4$

b. $\frac{3}{8}x = -6$

# Dividing Fractions and Mixed Numbers

**WHAT YOU'LL LEARN**

• Divide fractions and mixed numbers.

**EXAMPLE** Divide by a Fraction

❶ Find $\frac{2}{3} \div \frac{4}{9}$. Write in simplest form.

$$\frac{2}{3} \div \frac{4}{9} = \frac{2}{3} \cdot \boxed{\phantom{xx}}$$

Multiply by the reciprocal $\frac{4}{9}$.

$$= \frac{\overset{1}{2}}{\underset{1}{3}} \cdot \frac{\overset{3}{9}}{\underset{2}{4}}$$

Divide out common factors.

$$= \boxed{\phantom{xx}} \text{ or } \boxed{\phantom{xx}}$$

Multiply and simplify.

**KEY CONCEPT**

**Division by a Fraction**
To divide by a fraction, multiply by its multiplicative inverse or reciprocal.

**EXAMPLE** Divide by Mixed Numbers

❷ **JETS** An Air Force unit has 34 million dollars to spend on new jets. Each jet costs $4\frac{1}{4}$ million dollars. How many jets can be bought?

$$34 \div 4\frac{1}{4} = 34 \div \boxed{\phantom{xx}}$$

Rename $4\frac{1}{4}$ as an improper fraction.

$$= 34 \cdot \boxed{\phantom{xx}}$$

Multiply by the reciprocal of $\frac{17}{4}$.

$$= \frac{\overset{2}{34}}{1} \cdot \frac{4}{\underset{1}{17}}$$

Divide out common factors.

$$= \boxed{\phantom{xx}}$$

Multiply.

The Air Force unit can buy $\boxed{\phantom{x}}$ jets.

**Your Turn**

a. Find $\frac{6}{7} \div \frac{2}{5}$. Write in simplest form.

b. Linda has 22 cups of flour. She is using a cookie recipe that requires $2\frac{3}{4}$ cups of flour per batch. How many batches of cookies can Linda make?

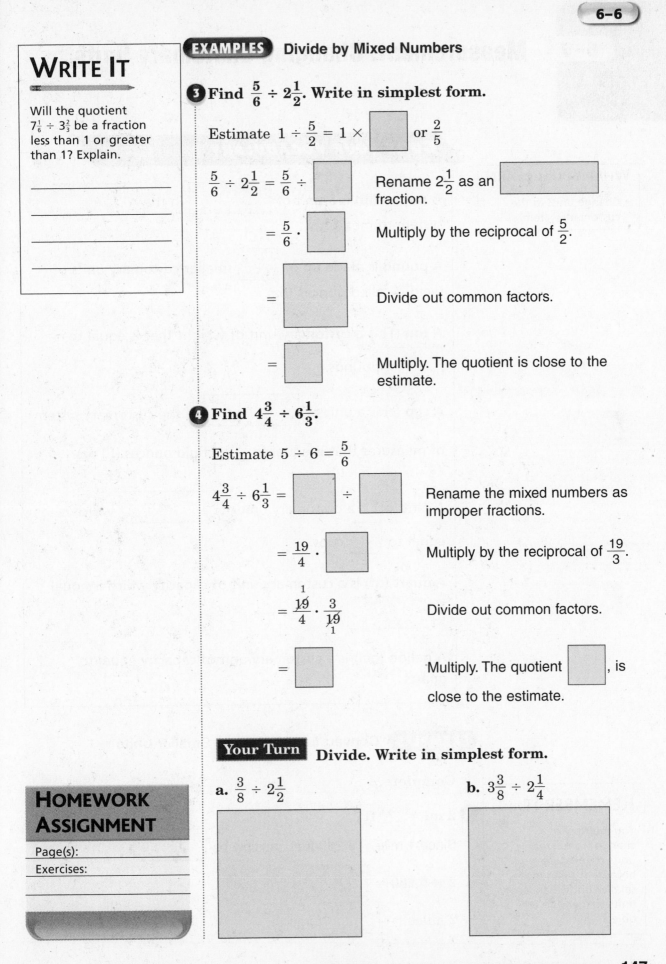

## WRITE IT

Will the quotient $7\frac{1}{6} \div 3\frac{2}{3}$ be a fraction less than 1 or greater than 1? Explain.

_____

_____

_____

_____

**EXAMPLES** **Divide by Mixed Numbers**

**❸ Find $\frac{5}{6} \div 2\frac{1}{2}$. Write in simplest form.**

Estimate $1 \div \frac{5}{2} = 1 \times \boxed{\phantom{x}}$ or $\frac{2}{5}$

$\frac{5}{6} \div 2\frac{1}{2} = \frac{5}{6} \div \boxed{\phantom{x}}$     Rename $2\frac{1}{2}$ as an $\boxed{\phantom{xxxx}}$ fraction.

$= \frac{5}{6} \cdot \boxed{\phantom{x}}$     Multiply by the reciprocal of $\frac{5}{2}$.

$= \boxed{\phantom{xx}}$     Divide out common factors.

$= \boxed{\phantom{x}}$     Multiply. The quotient is close to the estimate.

**❹ Find $4\frac{3}{4} \div 6\frac{1}{3}$.**

Estimate $5 \div 6 = \frac{5}{6}$

$4\frac{3}{4} \div 6\frac{1}{3} = \boxed{\phantom{x}} \div \boxed{\phantom{x}}$     Rename the mixed numbers as improper fractions.

$= \frac{19}{4} \cdot \boxed{\phantom{x}}$     Multiply by the reciprocal of $\frac{19}{3}$.

$= \frac{\overset{1}{\cancel{19}}}{4} \cdot \frac{3}{\underset{1}{\cancel{19}}}$     Divide out common factors.

$= \boxed{\phantom{x}}$     Multiply. The quotient $\boxed{\phantom{x}}$, is close to the estimate.

**Your Turn** **Divide. Write in simplest form.**

**a.** $\frac{3}{8} \div 2\frac{1}{2}$

**b.** $3\frac{3}{8} \div 2\frac{1}{4}$

## HOMEWORK ASSIGNMENT

Page(s): _____

Exercises: _____

_____

_____

*Mathematics: Applications and Concepts, Course 2*     **147**

## 6–7  Measurement: Changing Customary Units

**WHAT YOU'LL LEARN**

• Change units in the customary system.

**BUILD YOUR VOCABULARY** (pages 134–135)

A **pound** (lb) is a unit of [ ] in the [ ] system of measures.

A pound is made up of [ ] smaller customary units of weight called **ounces** (oz).

A **ton** (T) is a customary unit of weight that is equal to [ ] pounds.

A **cup** (c) is a unit of [ ] in the customary system of measures that is equal to [ ] fluid ounces (fl. oz).

A **pint** (pt) is a customary unit of [ ] which is equal to [ ] cups.

A **quart** (qt) is a customary unit of capacity which is equal to 2 [ ].

A **gallon** (gal) is a customary unit of capacity equal to [ ] quarts.

**EXAMPLES** Convert Larger Units to Smaller Units

**REMEMBER IT**

You multiply to change from larger units of measure because it takes more smaller units than larger units to measure an object.

Complete.

**1** 2 mi = __?__ ft

Since 1 mile = 5,280 feet, multiply by [ ].

2 × 5,280 = [ ]

2 miles = [ ] feet

**2** 5 pt = __?__ c

Since 1 pint = 2 cups, multiply by ▢.

5 × 2 = ▢

5 pints = ▢ cups

**REVIEW IT**

Explain how estimating can help you solve a problem. *(Lesson 6-3a)*

_____

_____

_____

_____

_____

_____

**Your Turn** Complete.

**a.** 8 yd = ? ft

**b.** 3 gal = ? pt

**EXAMPLE** Convert Units to Solve a Problem

**3** ELEVATOR The elevator in an office building has a weight limit posted of one and a half tons. How many pounds can the elevator safely hold?

$1\frac{1}{2} \times 2,000 =$ ▢    Multiply by ▢ since there

are ▢ pounds in 1 ton.

So the elevator can safely hold ▢ pounds.

**Your Turn** The label on a bag of potatoes states that the bag contains $3\frac{3}{4}$ pounds of potatoes. How many ounces of potatoes does the bag hold?

**EXAMPLES** Smaller Units to Larger Units

**4** Complete.

**1,000 in. = __?__ ft**

Since there are ▢ inches in 1 foot, divide by ▢.

$1,000 \div 12 =$ ▢ or ▢

$1,000$ inches $= 83\frac{1}{3}$ feet

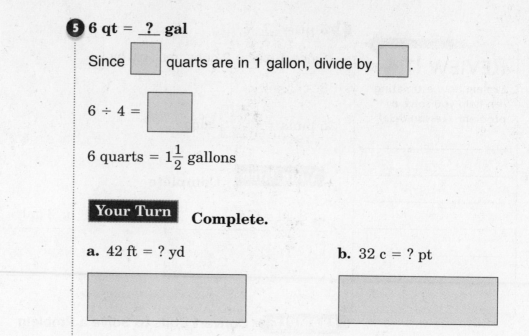

**⑤** 6 qt = __?__ gal

Since ☐ quarts are in 1 gallon, divide by ☐ .

6 ÷ 4 = ☐

6 quarts = $1\frac{1}{2}$ gallons

**Your Turn** Complete.

**a.** 42 ft = ? yd

**b.** 32 c = ? pt

# Geometry: Perimeter and Area

**BUILD YOUR VOCABULARY** page (135)

The ⬚ around a geometric figure is called a **perimeter**.

**EXAMPLE** Find the Perimeter of a Rectangle

❶ Find the perimeter of the figure.   ⬚ 2 ft
                                        18 ft

**KEY CONCEPT**

**Perimeter of a Rectangle**
The perimeter $P$ of a rectangle is twice the sum of the length $\ell$ and width $w$.

$P = 2\ell + 2w$          Perimeter of a rectangle

$P = 2(18) + 2(2)$       $\ell = $ ⬚ , $w = $ ⬚

$P = $ ⬚ $ + $ ⬚       Multiply.

$P = $ ⬚                   Add.

The perimeter is 40 ⬚ .

**EXAMPLE** Find the Perimeter of an Irregular Figure

❷ Find the perimeter of the figure.

$P = 2\frac{1}{4} + \frac{3}{4} + \frac{3}{4} + 2\frac{1}{8} + 1\frac{3}{8}$

$P = $ ⬚ $ + $ ⬚ $ + $ ⬚ $ + 2\frac{1}{8} + 1\frac{3}{8}$

$P = $ ⬚ or ⬚

The perimeter is ⬚ inches.

$\frac{3}{4}$ in. $\frac{3}{4}$ in.

$2\frac{1}{4}$ in.          $2\frac{1}{8}$ in.

$1\frac{3}{8}$ in.

**Your Turn** Find the perimeter of each figure.

a.              9 in.

14 in.

b.    2.7 ft      3.5 ft

3.8 ft

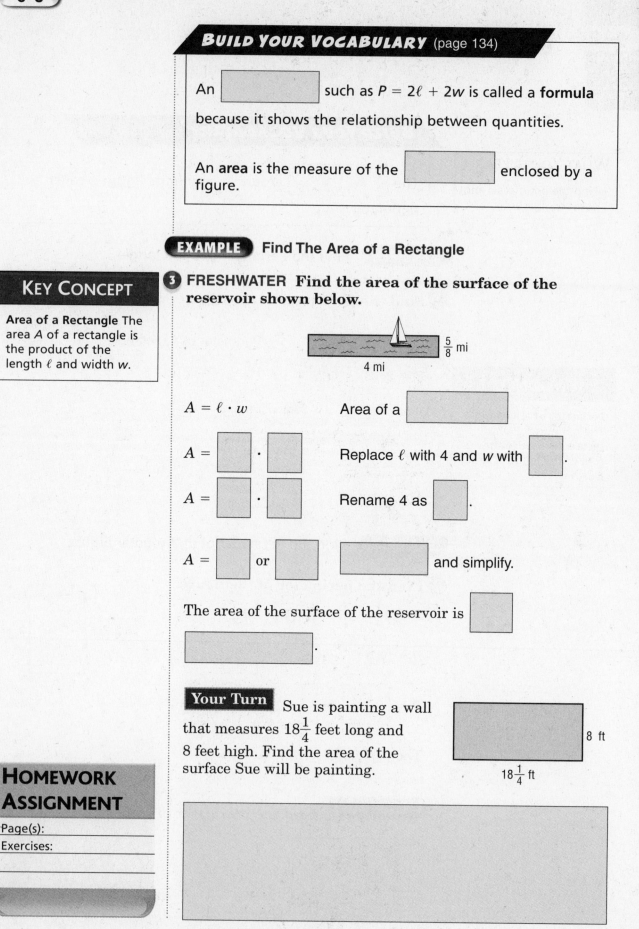

## BUILD YOUR VOCABULARY (page 134)

An [          ] such as $P = 2\ell + 2w$ is called a **formula** because it shows the relationship between quantities.

An **area** is the measure of the [          ] enclosed by a figure.

**EXAMPLE** Find The Area of a Rectangle

**KEY CONCEPT**

**Area of a Rectangle** The area $A$ of a rectangle is the product of the length $\ell$ and width $w$.

**3** FRESHWATER Find the area of the surface of the reservoir shown below.

$\frac{5}{8}$ mi

4 mi

$A = \ell \cdot w$          Area of a [          ]

$A = $ [   ] $\cdot$ [   ]          Replace $\ell$ with 4 and $w$ with [   ].

$A = $ [   ] $\cdot$ [   ]          Rename 4 as [   ].

$A = $ [   ] or [   ] [          ] and simplify.

The area of the surface of the reservoir is [   ] [          ].

**Your Turn** Sue is painting a wall that measures $18\frac{1}{4}$ feet long and 8 feet high. Find the area of the surface Sue will be painting.

8 ft

$18\frac{1}{4}$ ft

**HOMEWORK ASSIGNMENT**

Page(s):

Exercises:

# Geometry: Circles and Circumference

## WHAT YOU'LL LEARN

- Find the circumference of circles.

A **circle** is a set of all points in a plane that are the [    ] distance from a given [    ].

The set of all points in a [    ] are equi-distant from a [    ] called the **center**.

The **diameter** (d) is the distance [    ] a [    ] through its center.

The **circumference** (C) is the distance [    ] a circle.

The **radius** (r) is the distance from the [    ] to any point on a [    ].

---

**EXAMPLE**   Find Circumference

## KEY CONCEPT

**Circumference of a Circle**
The circumference C of a circle is equal to its diameter d times π, or 2 times its radius r times π.

**①PETS** Find the circumference around the hamster's running wheel shown. Round to the nearest tenth.

$C = 2\pi r$

$C = 2$ [    ] (3)

$C = $ [    ]          Multiply.

The circumference is about [    ] inches.

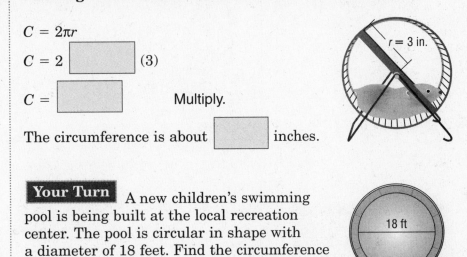

r = 3 in.

**Your Turn**   A new children's swimming pool is being built at the local recreation center. The pool is circular in shape with a diameter of 18 feet. Find the circumference of the pool. Round to the nearest tenth.

18 ft

## EXAMPLE Find Circumference

**2** Find the circumference of a circle with a diameter of 49 centimeters.

Since 49 is a multiple of 7, use [   ] for π.

$C = \pi d$  Circumference of a circle

$C \approx \dfrac{22}{7} \cdot$ [   ]  Replace [   ] with $\dfrac{22}{7}$ and $d$ with [   ].

$C \approx \dfrac{22}{7} \cdot \dfrac{\overset{7}{49}}{\underset{1}{1}}$  Divide by the [   ], 7.

$C \approx$ [   ]  Multiply.

The circumference is about 154 [   ].

**Your Turn** The circumference of a circle with a radius of 35 feet.

# BRINGING IT ALL TOGETHER

## STUDY GUIDE

| FOLDABLES™ | VOCABULARY PUZZLEMAKER | BUILD YOUR VOCABULARY |
|---|---|---|
| Use your **Chapter 6 Foldable** to help you study for your chapter test. | To make a crossword puzzle, word search, or jumble puzzle of the vocabulary words in Chapter 6, go to:<br><br>www.glencoe.com/sec/math/<br>t_resources/free/index.php | You can use your completed **Vocabulary Builder** *(pages 134–135)* to help you solve the puzzle. |

### 6-1

### Estimating with Fractions

**Which operation does math word indicate?**

1. sum

2. difference

3. product

4. quotient

**Estimate.**

5. $8\frac{2}{3} + 7\frac{1}{4}$

6. $11\frac{7}{8} \div 3\frac{5}{6}$

### 6-2

### Adding and Subtracting Fractions

**Choose the correct term to complete each sentence.**

7. To add or subtract fractions, add or subtract the (numerator, denominators) and write the results over the (numerator, denominators).

8. Unlike fractions have different (numerators, denominators).

9. The LCD is the (Least Common Denominator, Least Common Multiple).

**Add or subtract. Write in simplest form.**

10. $\frac{7}{8} + \frac{3}{8}$

11. $\frac{5}{6} - \frac{1}{3}$

12. $\frac{1}{5} + \frac{3}{4}$

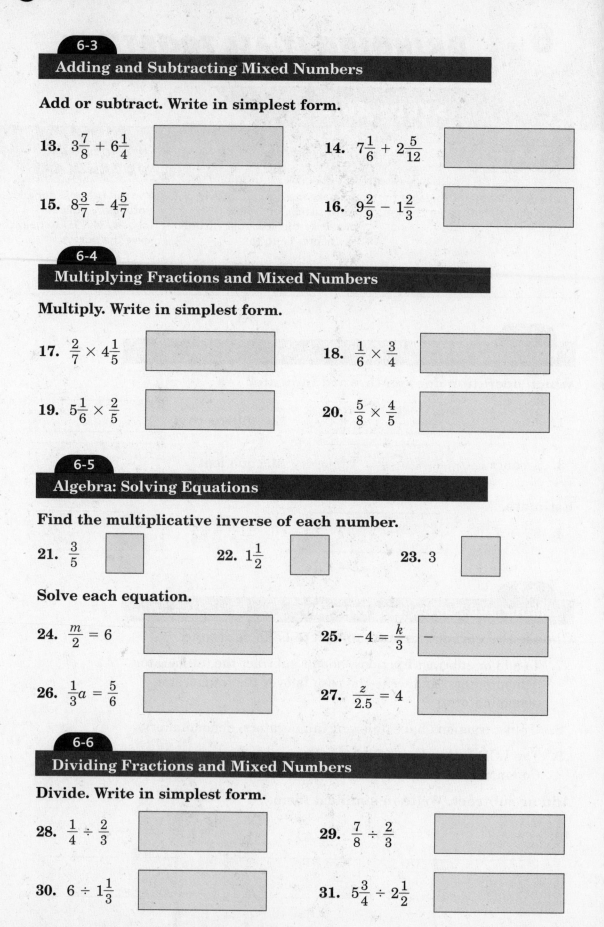

**6-3**

## Adding and Subtracting Mixed Numbers

**Add or subtract. Write in simplest form.**

13. $3\frac{7}{8} + 6\frac{1}{4}$

14. $7\frac{1}{6} + 2\frac{5}{12}$

15. $8\frac{3}{7} - 4\frac{5}{7}$

16. $9\frac{2}{9} - 1\frac{2}{3}$

**6-4**

## Multiplying Fractions and Mixed Numbers

**Multiply. Write in simplest form.**

17. $\frac{2}{7} \times 4\frac{1}{5}$

18. $\frac{1}{6} \times \frac{3}{4}$

19. $5\frac{1}{6} \times \frac{2}{5}$

20. $\frac{5}{8} \times \frac{4}{5}$

**6-5**

## Algebra: Solving Equations

**Find the multiplicative inverse of each number.**

21. $\frac{3}{5}$

22. $1\frac{1}{2}$

23. $3$

**Solve each equation.**

24. $\frac{m}{2} = 6$

25. $-4 = \frac{k}{3}$

26. $\frac{1}{3}a = \frac{5}{6}$

27. $\frac{z}{2.5} = 4$

**6-6**

## Dividing Fractions and Mixed Numbers

**Divide. Write in simplest form.**

28. $\frac{1}{4} \div \frac{2}{3}$

29. $\frac{7}{8} \div \frac{2}{3}$

30. $6 \div 1\frac{1}{3}$

31. $5\frac{3}{4} \div 2\frac{1}{2}$

**6-7**

## Measurement: Changing Customary Units

**To the right of each customary unit, write its abbreviation.**

**32.** inch

**33.** mile

**34.** ton

**35.** foot

**36.** ounce

**37.** quart

**Complete.**

**38.** $3\frac{3}{4}$ pt = ? c

**39.** 90 ft = ? yd

**40.** 156 oz = ? lb

**6-8**

## Geometry: Perimeter and Area

**41.** To find the perimeter of a _____ , add _____ the length to twice the _____ .

**Find the perimeter and area of each rectangle.**

**42.** $\ell = 6$ ft, $w = 3$ ft

**43.** $\ell = 3.4$ m, w = 5 m

**6-9**

## Geometry: Circles and Circumference

**Find the circumference of each circle. Use 3.14 or $\frac{22}{7}$ for $\pi$. Round to the nearest tenth if necessary.**

**44.** radius = 7.4 cm

**45.** radius = $3\frac{1}{2}$ in.

**46.** diameter = $6\frac{1}{8}$ ft

**47.** diameter = 1.7 mi

## CHAPTER 6 Checklist

# ARE YOU READY FOR THE CHAPTER TEST?

Visit **msmath2.net** to access your textbook, more examples, self-check quizzes, and practice tests to help you study the concepts in Chapter 6.

Check the one that applies. Suggestions to help you study are given with each item.

☐ **I completed the review of all or most lessons without using my notes or asking for help.**

- You are probably ready for the Chapter Test.

- You may want to take the Chapter 6 Practice Test on page 281 of your textbook as a final check.

☐ **I used my Foldable or Study Notebook to complete the review of all or most lessons.**

- You should complete the Chapter 6 Study Guide and Review on pages 278–280 of your textbook.

- If you are unsure of any concepts or skills, refer back to the specific lesson(s).

- You may also want to take the Chapter 6 Practice Test on page 281 of your textbook.

☐ **I asked for help from someone else to complete the review of all or most lessons.**

- You should review the examples and concepts in your Study Notebook and Chapter 6 Foldable.

- Then complete the Chapter 6 Study Guide and Review on pages 278–280 of your textbook.

- If you are unsure of any concepts or skills, refer back to the specific lesson(s).

- You may also want to take the Chapter 6 Practice Test on page 281 of your textbook.

|  |  |
|---|---|
| Student Signature | Parent/Guardian Signature |

Teacher Signature

# 7 CHAPTER

# Ratios and Proportions

**FOLDABLES™** Use the instructions below to make a Foldable to help you organize your notes as you study the chapter. You will see Foldable reminders in the margin of this Interactive Study Notebook to help you in taking notes.

---

**Begin with a sheet of notebook paper.**

**STEP 1** **Fold**
Fold lengthwise to the holes.

**STEP 2** **Cut**
Cut along the top line and then make equal cuts to form 10 tabs.

**STEP 3** **Label**
Label the major topics as shown.

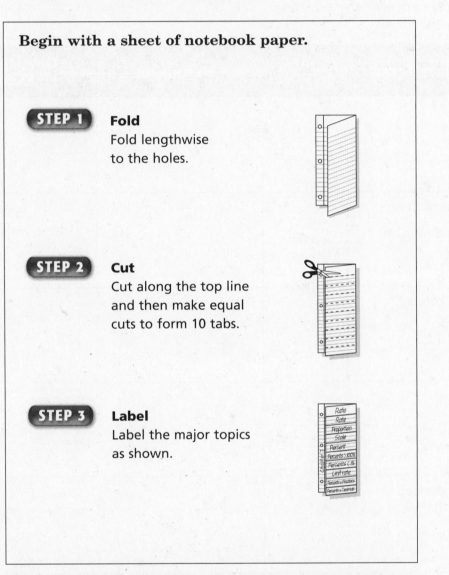

---

 **NOTE-TAKING TIP:** When you take notes, it may be helpful to include an example for each term or concept learned.

**Chapter 7**

# BUILD YOUR VOCABULARY

This is an alphabetical list of new vocabulary terms you will learn in Chapter 7. As you complete the study notes for the chapter, you will see Build Your Vocabulary reminders to complete each term's definition or description on these pages. Remember to add the textbook page number in the second column for reference when you study.

| Vocabulary Term | Found on Page | Definition | Description or Example |
|---|---|---|---|
| base | | | |
| cross products | | | |
| equivalent ratios | | | |
| part | | | |
| percent proportion | | | |
| proportion | | | |

| Vocabulary Term | Found on Page | Definition | Description or Example |
|---|---|---|---|
| rate | | | |
| scale | | | |
| scale drawing | | | |
| scale factor | | | |
| scale model | | | |
| unit rate | | | |

## WHAT YOU'LL LEARN

- Write ratios as fractions and determine whether two ratios are equivalent.

**FOLDABLES**

## ORGANIZE IT

Record a term or concept from Lesson 7–1 under the Ratios tab and write a definition along with an example to the right of the definition.

Ratio
Rate
Proportion
Scale
Percent
Percents > 100%
Percents < 1%
Unit rate
Percents & Fractions
Percents & Decimals

**EXAMPLES** **Write Ratios in Simplest Form**

**Write each ratio as a fraction in simplest form.**

**1** **30 to 9**

30 to 9 = [ ]          Write the ratio as a [ ].

= [ ]          Simplify.

Written as a fraction in simplest form,

the ratio 30 to 9 is [ ].

**2** **4:24**

4:24 = [ ]          Write the ratio as a [ ].

= [ ]          Simplify.

**Your Turn** **Write each ratio as a fraction in simplest form.**

**a.** 35 to 20                    **b.** 9:36

[ ]                              [ ]

**EXAMPLE** **Write Ratios by Converting Units**

**3** **Write the ratio 3 feet to 8 inches as a fraction in simplest form.**

$$\frac{3 \text{ feet}}{8 \text{ inches}} = \frac{[\qquad]}{8 \text{ inches}}$$          Convert feet to [ ].

$$= \frac{\overset{9}{\cancel{36 \text{ inches}}}}{\underset{2}{\cancel{8 \text{ inches}}}}$$          Divide by the [ ], 4 inches.

$$= [\quad]$$          Simplify.

**Your Turn** Write the ratio 4 feet to 20 inches as a fraction in simplest form.

---

**BUILD YOUR VOCABULARY** (page 154)

Two [  ] that have the same [  ] are equivalent ratios.

---

**EXAMPLE**  Compare Ratios

④ **Determine whether 12:15 and 32:40 are equivalent.**

Write each ratio as a fraction in simplest form.

$$12:15 = \frac{12 \div \boxed{\phantom{x}}}{15 \div \boxed{\phantom{x}}} \text{ or } \boxed{\phantom{x}}$$   The GCF of 12 and 15 is [  ].

$$32:40 = \frac{32 \div \boxed{\phantom{x}}}{40 \div \boxed{\phantom{x}}} \text{ or } \boxed{\phantom{x}}$$   The GCF of [  ] and [  ] is 8.

The ratios in simplest form both equal [  ]. So, 12:15 and

32:40 are [  ] ratios.

**REMEMBER IT**

Ratios such as 6:18 can also be written in simplest form as 1:3.

**Your Turn** Determine whether 8:24 and 14:42 are equivalent.

**HOMEWORK ASSIGNMENT**

Page(s): _____

Exercises: _____

# Rates

**WHAT YOU'LL LEARN**

- Determine unit rates.

A ratio that [          ] two quantities with different kinds of units is called a **rate**.

When a rate is simplified so that it has a [          ] of 1 unit, it is called a **unit rate**.

**FOLDABLES**

## ORGANIZE IT

Under the rate tab, take notes on rate and unit rate. Be sure to include examples.

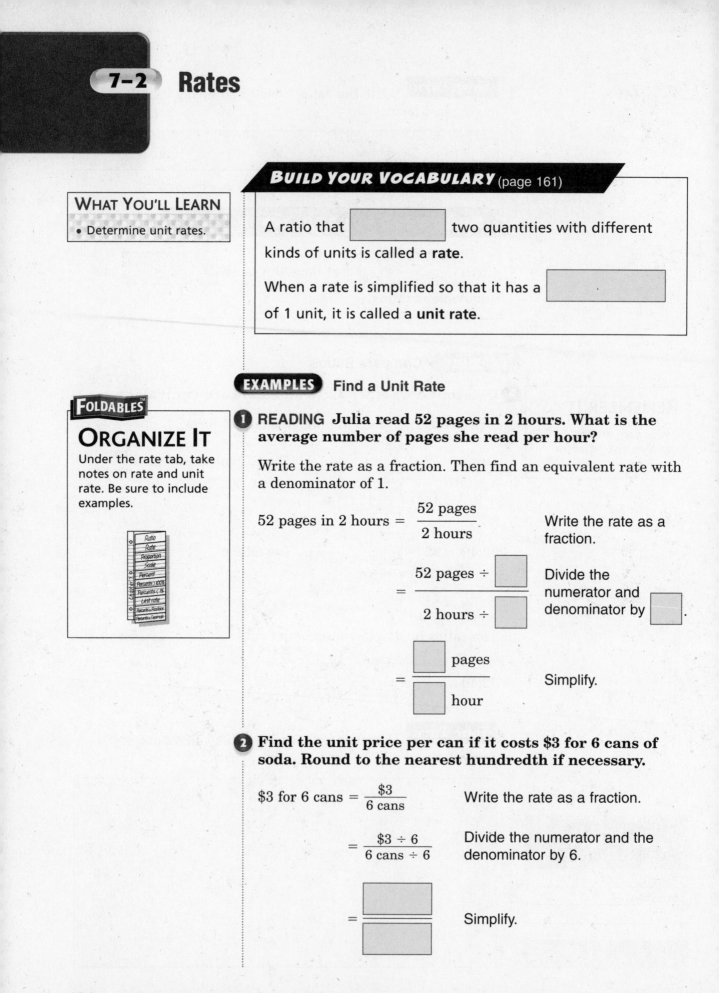

**EXAMPLES** Find a Unit Rate

**1** READING Julia read 52 pages in 2 hours. What is the average number of pages she read per hour?

Write the rate as a fraction. Then find an equivalent rate with a denominator of 1.

52 pages in 2 hours $= \dfrac{52 \text{ pages}}{2 \text{ hours}}$    Write the rate as a fraction.

$= \dfrac{52 \text{ pages} \div \boxed{\phantom{x}}}{2 \text{ hours} \div \boxed{\phantom{x}}}$    Divide the numerator and denominator by $\boxed{\phantom{x}}$.

$= \dfrac{\boxed{\phantom{x}} \text{ pages}}{\boxed{\phantom{x}} \text{ hour}}$    Simplify.

**2** Find the unit price per can if it costs $3 for 6 cans of soda. Round to the nearest hundredth if necessary.

$3 for 6 cans $= \dfrac{\$3}{6 \text{ cans}}$    Write the rate as a fraction.

$= \dfrac{\$3 \div 6}{6 \text{ cans} \div 6}$    Divide the numerator and the denominator by 6.

$= \dfrac{\boxed{\phantom{xxx}}}{\boxed{\phantom{xxx}}}$    Simplify.

**Your Turn**

**a.** Kyle skated 16 laps around the ice rink in 4 minutes. What is the average number of laps he skated per minute?

**b.** Find the unit price per cookie if it costs $3 for one dozen cookies. Round to the nearest hundredth if necessary.

**EXAMPLE** Choose the Best Buy

**3** The costs of different sizes of orange juice are shown in the table. Which container costs the least per ounce?

| Amount | Total Cost |
| --- | --- |
| 16 oz | $1.28 |
| 32 oz | $1.92 |
| 64 oz | $2.56 |
| 96 oz | $3.36 |

Find the unit price, or the cost per ounce of each size of orange juice. Divide the price by the number of ounces.

$1.28 ÷ _____ ounces = _____ per ounce.

$1.92 ÷ _____ ounces = _____ per ounce.

$2.56 ÷ _____ ounces = _____ per ounce.

$3.36 ÷ _____ ounces = _____ per ounce.

The _____-ounce container of orange juice costs the least per ounce.

**Your Turn** The costs of different sizes of bottles of laundry detergent are shown below. Which bottle costs the least per ounce?

| Amount | Total Cost |
| --- | --- |
| 16 oz | $3.12 |
| 32 oz | $5.04 |
| 64 oz | $7.04 |
| 96 oz | $11.52 |

# 7-3 Solving Proportions

## WHAT YOU'LL LEARN
- Solve proportions.

**BUILD YOUR VOCABULARY** (page 160)

When two ratios are [          ], they form a **proportion**.

In a proportion, a **cross-product** is the [          ] of the numerator of one ratio and the denominator of the other ratio.

## KEY CONCEPT

**Proportion** A proportion is an equation stating that two ratios are equivalent.

**EXAMPLES** Identify a Proportion

**Determine whether each pair of fractions forms a proportion.**

**1** $\frac{8}{12}$ and $\frac{2}{3}$

$$\frac{8}{12} \stackrel{?}{=} \frac{2}{3}$$  Write a proportion.

$$[\quad] \stackrel{?}{=} [\quad]$$  Find the [                    ].

$$24 = 24 \checkmark$$  [                ].

The cross products are [          ], so the ratios form a [              ].

**2** $\frac{\$5}{8 \text{ oz}}$ and $\frac{\$18}{32 \text{ oz}}$

$$\frac{5}{8} \stackrel{?}{=} \frac{18}{32}$$  Write a proportion.

$$[\quad] \stackrel{?}{=} [\quad]$$  Find the cross products.

$$160 \,[\quad]\, 144$$  Multiply.

The cross products are [          ], so the ratios [      ] form a proportion.

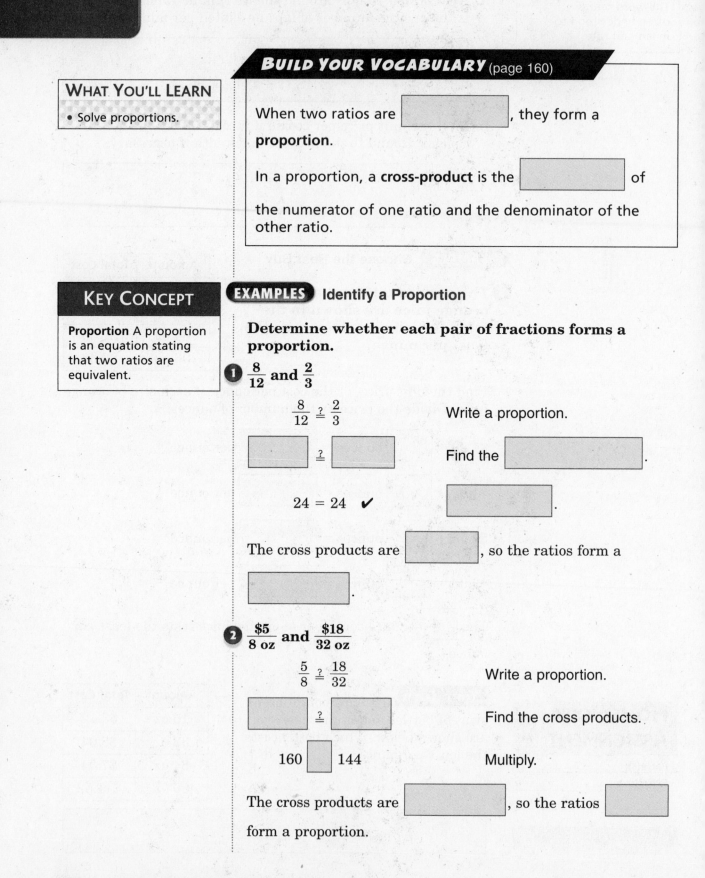

**EXAMPLE** Solve Proportions

③ Solve $\frac{3.5}{14} = \frac{6}{n}$.

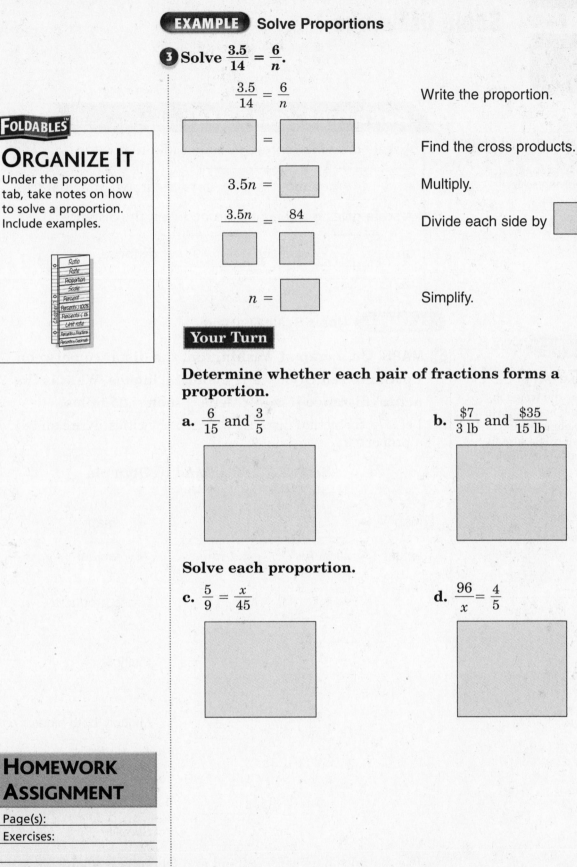

$$\frac{3.5}{14} = \frac{6}{n}$$     Write the proportion.

☐ = ☐     Find the cross products.

$3.5n = $ ☐     Multiply.

$\frac{3.5n}{☐} = \frac{84}{☐}$     Divide each side by ☐.

$n = $ ☐     Simplify.

**Your Turn**

**Determine whether each pair of fractions forms a proportion.**

**a.** $\frac{6}{15}$ and $\frac{3}{5}$

**b.** $\frac{\$7}{3\text{ lb}}$ and $\frac{\$35}{15\text{ lb}}$

**Solve each proportion.**

**c.** $\frac{5}{9} = \frac{x}{45}$

**d.** $\frac{96}{x} = \frac{4}{5}$

**HOMEWORK ASSIGNMENT**

Page(s):
Exercises:

**Scale Drawings**

### BUILD YOUR VOCABULARY (page 161)

A **scale drawing** represents something that is too

[         ] or too [         ] to be drawn at actual size.

A **scale** gives the relationship between the distance

on a [         ] and the [         ] distance.

---

### EXAMPLE  Use a Scale Drawing

1  **MAPS** On a map of Washington, the distance between Portland and Olympia is about $1\frac{7}{8}$ inches. What is the actual distance if the scale is $\frac{3}{8}$ inch = 25 miles?

Let $d$ = the actual distance between the cities. Write and solve a proportion.

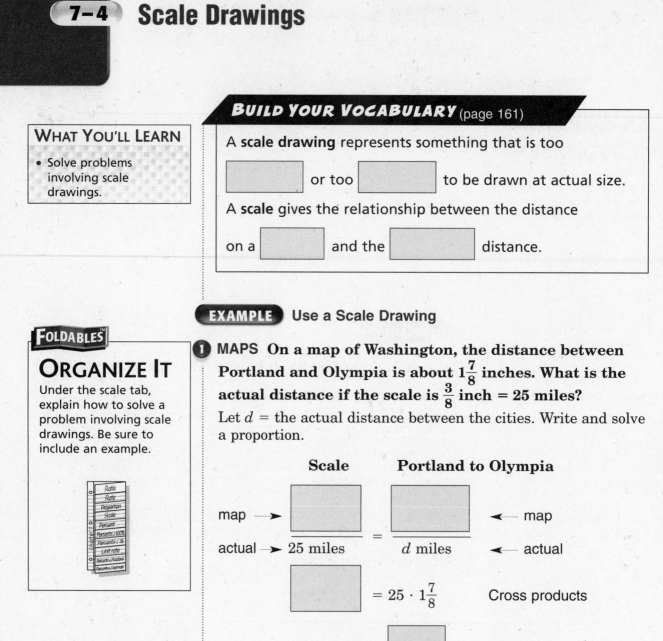

| Scale | Portland to Olympia |
|---|---|

map → [         ] = [         ] ← map

actual → 25 miles   $d$ miles   ← actual

[         ] $= 25 \cdot 1\frac{7}{8}$    Cross products

$\frac{3}{8}d =$ [         ]    Multiply.

[         ] $\cdot \frac{3}{8}d =$ [         ] $\cdot \frac{8}{3}$    Multiply both sides

by [         ].

$d =$ [         ]    Simplify.

The distance between Portland and Olympia is about

[         ].

### EXAMPLE  Read a Scale Drawing

**2 ARCHITECTURE** On the blueprint of a new house, each square has a side length of $\frac{1}{4}$ inch. If the length of a bedroom on the blueprint is $1\frac{1}{2}$ inches, what is the actual length of the room?

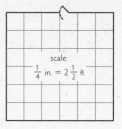

Write and solve a proportion.

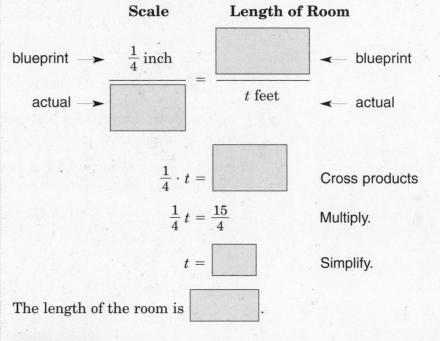

The length of the room is ⬚ .

---

**Your Turn**

**a.** On a map, the scale is given as 2 inches = 100 miles. If the distance on the map between the two cities is 15 inches, what is the actual distance between the two cities?

**b.** On a blueprint of a new house, each square has a side length of $\frac{1}{4}$ inch. If the width of the kitchen on the blueprint is 2 inches, what is the actual width of the room?

scale:
$\frac{1}{4}$ in. = 3 ft

---

## WRITE IT

Explain why these two scales are equivalent scales:

$\frac{1}{2}$ inch = 4 miles

1 inch = 8 miles

_____

_____

_____

---

## BUILD YOUR VOCABULARY (page 161)

A scale written as a [____] in [____] form is called the **scale factor**.

A **scale model** can be used to [____] something that is too large or too small for an actual-size [____].

**EXAMPLE** Find the Scale Factor

**3** Find the scale factor of a blueprint if the scale is $\frac{1}{2}$ inch = 3 feet.

Write the ratio of $\frac{1}{2}$ inch to 3 feet in simplest form.

$$\frac{\frac{1}{2} \text{ inch}}{3 \text{ feet}} = \frac{\frac{1}{2} \text{ inch}}{[\quad]}$$  Convert 3 feet to [____]

$$= [\quad] \cdot \frac{\frac{1}{2} \text{ inch}}{36 \text{ inches}}$$  Multiply by [____] to eliminate the fraction in the numerator.

$$= [\quad]$$  Cancel the units.

The scale factor is [____]. That is, each measure on the

blueprint is [____] the [____] measure.

**Your Turn** Find the scale factor of a blueprint if the scale is 1 inch = 4 feet.

[____]

## HOMEWORK ASSIGNMENT

Page(s):

Exercises:

# Fractions, Decimals, and Percents

Percents as Fractions

**WHAT YOU'LL LEARN**

- Write percents as fractions, and vice versa.

**FOLDABLES**

## ORGANIZE IT

Under the percent tab, take notes on writing percents as fractions and fractions as percents. Include examples.

Ratio
Rate
Proportion
Scale
Percent
Percents > 100%
Percents < 1%
Unit rate
Percents & Fractions
Percents & Decimals

**1** NUTRITION In a recent consumer poll, 41.8% of the people surveyed said they gained nutrition knowledge from family and friends. What fraction is this? Write in simplest form.

$41.8\% = \dfrac{41.8}{100}$    Write a fraction with a denominator of 100.

$= \dfrac{41.8}{100} \cdot \boxed{\phantom{x}}$    Multiply to eliminate the decimal in the numerator.

$= \boxed{\phantom{xx}}$  or  $\boxed{\phantom{xx}}$    Simplify.

**2** Write $12\frac{1}{2}\%$ as a fraction in simplest form.

$12\frac{1}{2}\% = \dfrac{12\frac{1}{2}}{100}$    Write a fraction.

$= 12\frac{1}{2} \div 100$    Divide.

$= \boxed{\phantom{xx}} \div 100$    Write $12\frac{1}{2}$ as an improper fraction.

$= \boxed{\phantom{xx}} \times \boxed{\phantom{xx}}$    Multiply by the reciprocal of 100.

$= \boxed{\phantom{xx}}$  or  $\boxed{\phantom{xx}}$    Simplify.

**Your Turn**

**a.** In a recent election, 64.8% of registered voters actually voted. What fraction is this? Write in simplest form.

**b.** Write $62\frac{1}{2}\%$ as a fraction in simplest form.

## KEY CONCEPTS

**Common Fraction/ Decimal/Percent Equivalents**

$\frac{1}{3} = 0.3 = 33\frac{1}{3}\%$

$\frac{2}{3} = 0.6 = 66\frac{2}{3}\%$

$\frac{1}{8} = 0.125 = 12\frac{1}{2}\%$

$\frac{3}{8} = 0.375 = 37\frac{1}{2}\%$

$\frac{5}{8} = 0.625 = 62\frac{1}{2}\%$

$\frac{7}{8} = 0.875 = 87\frac{1}{2}\%$

**EXAMPLES** Fractions as Percents

**Write each fraction as a percent. Round to the nearest hundredth if necessary.**

**3** $\frac{5}{12}$

To write the fraction as a percent, you can use a proportion.

$$\frac{5}{12} = \frac{n}{100}$$    Write a product using $\frac{n}{100}$.

$$\boxed{\phantom{xxx}} = \boxed{\phantom{xxx}}$$    Find the cross products.

$500 \boxed{\div} 12 \boxed{\text{ENTER}} \ 41.66666667$    Use a calculator.

So, $\frac{5}{12}$ is about $\boxed{\phantom{xxxx}}$.

**4** $\frac{9}{20}$

To write the fraction as a percent you can either use a proportion or first write the fraction as a decimal and then write the decimal as a percent.

$\frac{9}{20} = \boxed{\phantom{xxx}}$    Write $\frac{9}{20}$ as a decimal.

$= \boxed{\phantom{xxx}}$    Multiply by $\boxed{\phantom{x}}$ and add the $\boxed{\phantom{x}}$.

**5** $\frac{3}{7}$

$\frac{3}{7} = 0.4285714\ldots$    Write $\frac{3}{7}$ as a decimal.

$= \boxed{\phantom{xxx}}$    $\boxed{\phantom{xxxx}}$ by 100 and add the $\boxed{\phantom{x}}$.

**Your Turn** Write each fraction as a percent. Round to the nearest hundredth.

**a.** $\frac{11}{15}$

**b.** $\frac{13}{25}$

**c.** $\frac{9}{11}$

## HOMEWORK ASSIGNMENT

Page(s): _____

Exercises: _____

# Percents Greater Than 100% and Percents Less Than 1%

- Write percents greater than 100% and percents less than 1% as fractions and as decimals, and vice versa.

**EXAMPLES** Percents as Decimals or Fractions

**Write each percent as a decimal and as a mixed number or fraction in simplest form.**

**1** 220%

220% = [ ]          Definition of percent

= [ ]          Write as a decimal.

= [ ]          Write as a mixed number.

**FOLDABLES™**

## ORGANIZE IT

Under the percent > 100 tab, take notes on writing percents greater than 100% and percents less than 1%. Include examples.

Ratio
Rate
Proportion
Scale
Percent
Percents > 100%
Percents < 1%
Unit rate
Percents & Fractions
Percents & Decimals

**2** 0.6%

0.6% = [ ]          Definition of percent

= [ ]          Write as a decimal.

= [ ]

**3** STOCKS During a stock market rally, a company's stock increased in value by 200%. Write 200% as a decimal.

200% = 200          Divide by 100.

= [ ] or 2          Simplify.

**Your Turn** Write each percent as a decimal and as a mixed number or fraction in simplest form.

**a.** 375%

**b.** 0.4%

**c.** 420%

**EXAMPLES** Decimals as Percents

Write each decimal as a percent.

**4** **5.12**

$5.12 = 5.12$    Multiply by [ ] .

$= $ [ ]

**5** **0.0015**

$0.0015 = 0.0015$    Multiply by [ ] .

$= $ [ ]

**Your Turn** Write each decimal as a percent.

**a.** 0.0096

**b.** 9.35

**HOMEWORK ASSIGNMENT**

Page(s):

Exercises:

# Percent of a Number

**EXAMPLE** Use a Proportion to Find a Percent

**1** SURVEYS Out of 1,423 adults surveyed, 30% said they knew the name of their mail carrier. How many of the people surveyed knew their mail carrier's name?

30% means that 30 out of 100 people knew their mail carrier's name. Find an equivalent ratio $x$ out of 1,423 and write a proportion.

number that knew name $\rightarrow \dfrac{x}{1{,}423} = \dfrac{30}{100}$ percent of people that knew name
total number in survey

Now solve the proportion.

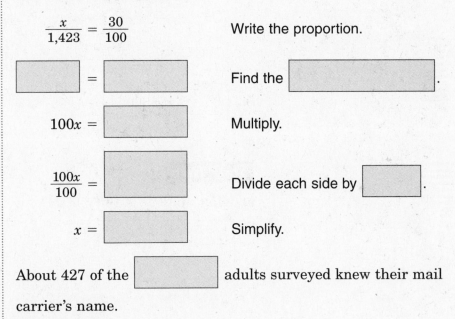

$$\frac{x}{1{,}423} = \frac{30}{100} \qquad \text{Write the proportion.}$$

$$\boxed{\phantom{xxx}} = \boxed{\phantom{xxx}} \qquad \text{Find the } \boxed{\phantom{xxxxxx}}.$$

$$100x = \boxed{\phantom{xxx}} \qquad \text{Multiply.}$$

$$\frac{100x}{100} = \boxed{\phantom{xxx}} \qquad \text{Divide each side by } \boxed{\phantom{xx}}.$$

$$x = \boxed{\phantom{xxx}} \qquad \text{Simplify.}$$

About 427 of the $\boxed{\phantom{xxxx}}$ adults surveyed knew their mail carrier's name.

**Your Turn** Out of 765 students surveyed, 42% said that they watch some television after school before doing their homework. How many of the students surveyed watch some television after school before doing their homework?

**EXAMPLES** Use Multiplication to Find a Percent

**2** **What number is 120% of 24?**

120% of 24 = 120% × 24      Write a multiplication expression.

= ☐ × 24      Write 120% as a decimal.

= ☐      Multiply.

So, 120% of ☐ is ☐.

**3** **Find 25% of $600.**

25% of $600 = 25% × $600      Write a multiplication expression.

= ☐ × 600      Write 25% as a decimal.

= ☐      Multiply.

So, ☐ of $600 is ☐.

**Your Turn**

**a.** What number is 160% of 44?

**b.** Find 80% of $450.

**HOMEWORK ASSIGNMENT**

Page(s):

Exercises:

# The Percent Proportion

### WHAT YOU'LL LEARN

- Solve problems using the percent proportion.

A **percent proportion** compares **part** of a quantity to the whole quantity, called the **base**, using a percent.

### KEY CONCEPT

**Percent Proportion** The percent proportion is $\frac{\text{part}}{\text{base}} = \frac{\text{percent}}{100}$.

**EXAMPLE** Find the Percent

**1** **What percent of 24 is 18?**

18 is the part, and 24 is the base. You need to find the percent.

$\frac{a}{b} = \frac{p}{100}$          Percent proportion

$\boxed{\phantom{xx}} = \frac{p}{100}$          $a = \boxed{\phantom{xx}}$, $b = \boxed{\phantom{xx}}$

$\phantom{xxx} = 24 \cdot p$          Find the cross products.

$1{,}800 = 24p$          Simplify.

$\boxed{\phantom{xx}} = \frac{24p}{24}$          Divide each side by $\boxed{\phantom{xx}}$.

$\boxed{\phantom{xx}} = p$          Simplify.

So, $\boxed{\phantom{xx}}$ of 24 is $\boxed{\phantom{xx}}$.

**EXAMPLE** Find the Part

**2** **What number is 30% of 150?**

30 is the percent and 150 is the base. You need to find the part.

$\frac{a}{b} = \frac{p}{100}$          Percent proportion

$\frac{a}{150} = \boxed{\phantom{xxx}}$          $b = \boxed{\phantom{xx}}$, $p = \boxed{\phantom{xx}}$

$a \cdot 100 = 150 \cdot 30$          Find the cross products.

$100a = \boxed{\phantom{xxx}}$          Simplify.

$\frac{100a}{100} = \frac{4{,}500}{100}$          Divide each side by 100.

$a = \boxed{\phantom{xx}}$          Simplify.

So, 30% of $\boxed{\phantom{xx}}$ is 45.

**EXAMPLE** Find the Base

**3** **12 is 80% of what number?**

12 is the part and 80 is the percent. You need to find the base.

$$\frac{a}{b} = \frac{p}{100}$$     Percent proportion

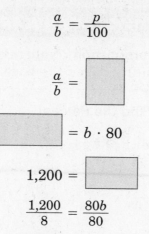

$$\frac{a}{b} = \boxed{\phantom{xx}}$$     $a = \boxed{\phantom{xx}}$, $p = 80$.

$$\boxed{\phantom{xxxx}} = b \cdot 80$$     Find the cross products.

$$1{,}200 = \boxed{\phantom{xxx}}$$     Simplify.

$$\frac{1{,}200}{8} = \frac{80b}{80}$$     Divide each side by $\boxed{\phantom{x}}$.

$$\boxed{\phantom{xx}} = b$$

So, 12 is 80% of 15.

**WRITE IT**

Write an example of a real-world percent problem.

_____

_____

_____

_____

**Your Turn**

**a.** What percent of 80 is 28?

**b.** What number is 65% of 180?

**c.** 36 is 40% of what number?

**HOMEWORK ASSIGNMENT**

Page(s): _____

Exercises: _____

_____

_____

# BRINGING IT ALL TOGETHER

## STUDY GUIDE

| FOLDABLES | VOCABULARY PUZZLEMAKER | BUILD YOUR VOCABULARY |
|---|---|---|
| Use your **Chapter 7 Foldable** to help you study for your chapter test. | To make a crossword puzzle, word search, or jumble puzzle of the vocabulary words in Chapter 7, go to:<br><br>www.glencoe.com/sec/math/t_resources/free/index.php | You can use your completed **Vocabulary Builder** *(pages 160–161)* to help you solve the puzzle. |

### Ratios

**State whether each sentence is true or false. If false, replace the underlined word to make it a true sentence.**

1. When you simplify a ratio, write a fraction as <u>a mixed number</u>.

2. To write a ratio comparing measures, both quantities should have <u>the same</u> unit of measure.

**Write each ratio as a fraction in simplest form.**

3. 63:7

4. 15:54

### Rates

**Complete.**

5. A [  ] is a ratio that compares two quantities with different kinds of units.

**Write each ratio as a fraction in simplest form.**

6. 36 inches: 48 inches

7. 15 minutes to 3 hours

*Mathematics: Applications and Concepts, Course 2*    **179**

## 7-3
## Solving Proportions

**Complete each sentence.**

8. The cross products of a [          ] are equal.

9. If you know [          ] parts of a proportion, you can solve for

   the fourth part by [                    ] and then

   [          ] both sides by the co-efficient of the unknown.

**Solve each proportion.**

10. $\frac{15}{n} = \frac{3}{8}$ [          ]

11. $\frac{6}{20} = \frac{x}{80}$ [          ]

12. $\frac{b}{16} = \frac{3}{48}$ [          ]

13. $\frac{35}{90} = \frac{14}{y}$ [          ]

## 7-4
## Scale Drawings

On a map, the scale is $\frac{1}{4}$ inch = 10 miles. For each map
distance, find the actual distance.

14. 6 inches [          ]

15. $\frac{3}{8}$ inch [          ]

16. $2\frac{1}{2}$ inches [          ]

17. 1 inch [          ]

## 7-5
## Fractions, Decimals, and Percents

**Complete the table of equivalent fractions.**

| | Fraction | Decimal | Percent |
|---|---|---|---|
| 18. | $\frac{1}{3}$ | [  ] | [  ] |
| 19. | $\frac{3}{8}$ | [  ] | $37\frac{1}{2}\%$ |
| 20. | $\frac{1}{8}$ | [  ] | [  ] |
| 21. | [  ] | 0.875 | $87\frac{1}{2}\%$ |

**7-6**

## Percents Greater Than 100% and Percents Less Than 1%

**Write each percent as a decimal and as a mixed number or fraction in simplest form.**

**22.** 150%

**23.** 0.25%

**Write each decimal as a percent.**

**24.** 2.75

**25.** 0.0043

**Write each number as a percent.**

**26.** $5\frac{1}{4}$

**27.** $\frac{4}{2000}$

**7-7**

## Percent of a Number

**Find each number.**

**28.** What is 3% of 530?

**29.** Find 15% of $24.

**30.** Find 200% of 17.

**31.** What is .6% of 800?

**7-8**

## The Percent Proportion

**32.** In the formula $\frac{a}{b} = \frac{p}{100}$, $a$ is the [            ], $b$ is the

[            ], and $p$ is the [            ].

**33.** What number is 30% of 15?

**34.** 32.5 is 65% of what number?

# ARE YOU READY FOR THE CHAPTER TEST?

Visit **msmath2.net** to access your textbook, more examples, self-check quizzes, and practice tests to help you study the concepts in Chapter 7.

Check the one that applies. Suggestions to help you study are given with each item.

☐ **I completed the review of all or most lessons without using my notes or asking for help.**

- You are probably ready for the Chapter Test.

- You may want to take the Chapter 7 Practice Test on page 329 of your textbook as a final check.

☐ **I used my Foldable or Study Notebook to complete the review of all or most lessons.**

- You should complete the Chapter 7 Study Guide and Review on pages 326–328 of your textbook.

- If you are unsure of any concepts or skills, refer back to the specific lesson(s).

- You may also want to take the Chapter 7 Practice Test on page 329 of your textbook.

☐ **I asked for help from someone else to complete the review of all or most lessons.**

- You should review the examples and concepts in your Study Notebook and Chapter 7 Foldable.

- Then complete the Chapter 7 Study Guide and Review on pages 326–328 of your textbook.

- If you are unsure of any concepts or skills, refer back to the specific lesson(s).

- You may also want to take the Chapter 7 Practice Test on page 329 of your textbook.

Student Signature

Parent/Guardian Signature

Teacher Signature

# Applying Percent

FOLDABLES™ Use the instructions below to make a Foldable to help you organize your notes as you study the chapter. You will see Foldable reminders in the margin of this Interactive Study Notebook to help you in taking notes.

**Begin with a piece of 11" × 17" paper.**

**STEP 1** **Fold**
Fold a 2" tab along the long side of the paper. Then fold the rest in half.

**STEP 2** **Open and Fold**
Open the paper and fold in half widthwise 3 times to make 8 columns.

**STEP 3** **Open and Label**
Draw lines along the folds and label as shown.

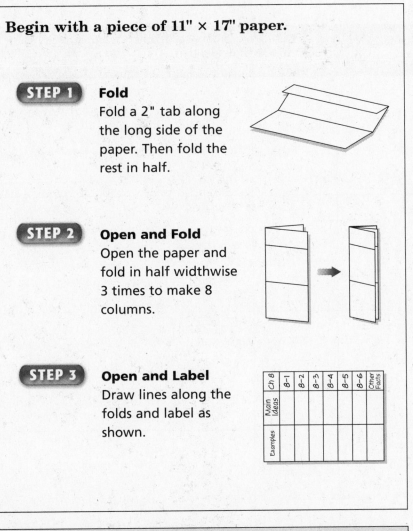

 NOTE-TAKING TIP: When you take notes, it is often helpful to reflect on ways the concepts apply to your daily life.

## BUILD YOUR VOCABULARY

This is an alphabetical list of new vocabulary terms you will learn in Chapter 8. As you complete the study notes for the chapter, you will see Build Your Vocabulary reminders to complete each term's definition or description on these pages. Remember to add the textbook page number in the second column for reference when you study.

| Vocabulary Term | Found on Page | Definition | Description or Example |
|---|---|---|---|
| discount | | | |
| percent equation | | | |
| percent of change | | | |
| percent of decrease | | | |
| percent of increase | | | |

| Vocabulary Term | Found on Page | Definition | Description or Example |
|---|---|---|---|
| population | | | |
| principal | | | |
| random sample | | | |
| sales tax | | | |
| simple interest | | | |
| survey | | | |

# Percent and Estimation

## WHAT YOU'LL LEARN

- Estimate percents by using fractions and decimals.

**EXAMPLES** Use Fractions to Estimate

**Estimate by using a fraction.**

**① 61% of 407**

61% is about ☐%, which is ☐ or ☐ .

61% of 407 ≈ ☐ · ☐ . Use ☐ to estimate and round 407 to ☐ .

≈ ☐ Multiply.

So, 61% of 407 is about ☐ .

**② 18% of 296**

18% is about ☐%, which is ☐ or ☐ .

18% of 296 ≈ ☐ · ☐ . Use ☐ to estimate and round 296 to ☐ .

≈ ☐ Multiply.

18% of 296 is about ☐ .

**EXAMPLE** Estimate Using 10%

**③ Estimate 59% of 500.**

**Step 1** Find 10% of the number.

10% of 500 = ☐ · 500   To multiply by 10%, move the decimal point one place to the ☐ .

= ☐

**Step 2**  Multiply.

59% is about [ ] %.

60% of 500 is [ ] times 10% of 500.

6 · 50 = [ ]

So, 59% of 500 is about [ ] .

**Your Turn**  Estimate 39% of 700.

[ ]

---

**EXAMPLE**  Percents Greater Than 100 or Less Than 1

**REMEMBER IT**

To estimate the percent of a number, round the percent, round the number, or round both.

④ **Estimate.**

**173% of 60**

173% is more than 100%, so 173% of 60 is greater than 60%.

173% is about [ ] .

175% of 60

= [ ]  +  [ ]         175% = 100% + 75%

= (1 · 60) + ( [ ] · 60)         100% = 1 and 75% = [ ]

= [ ]  or  [ ]              Simplify.

So, 173% of 60 is about [ ] .

**Your Turn**  **Estimate.**

**a.** 28% of 92

[ ]

**b.** 71% of 198

[ ]

**c.** 142% of 80

[ ]

**d.** $\frac{1}{5}$% of 1,002

[ ]

**HOMEWORK ASSIGNMENT**

Page(s):

Exercises:

# The Percent Equation

**BUILD YOUR VOCABULARY** (Page 184)

The equation [ ] = percent · [ ] is called the **percent equation**.

**FOLDABLES**

## ORGANIZE IT

Record the main ideas, and give examples about the percent equation in the row for Lesson 2 of your Foldable.

**EXAMPLE** Find the Part

**1** What number is 46% of 200?

46% or [ ] is the percent and [ ] is the base.

Let $n$ represent the [ ].

$$\underbrace{\text{part}} = \underbrace{\text{percent}} \cdot \underbrace{\text{base}}$$

$n$ = [ ] · 200   Write an equation.

$n$ = [ ]   Multiply.

So, 46% of 200 is [ ].

**EXAMPLE** Find the Percent

**2** 26 is what percent of 32?

Let $n$ represent the percent.

$$\underbrace{\text{part}} = \underbrace{\text{percent}} \cdot \underbrace{\text{base}}$$

[ ] = $n$ · 32   Write an equation.

[ ] = [ ]   Divide each side by [ ].

[ ] = $n$   Simplify.

[ ] = $n$   Write as a percent.

So, 26 is [ ] of 32.

**EXAMPLE** Find the Base

**3** **12 is 40% of what number?**

Let *n* represent the base.

part = percent · base

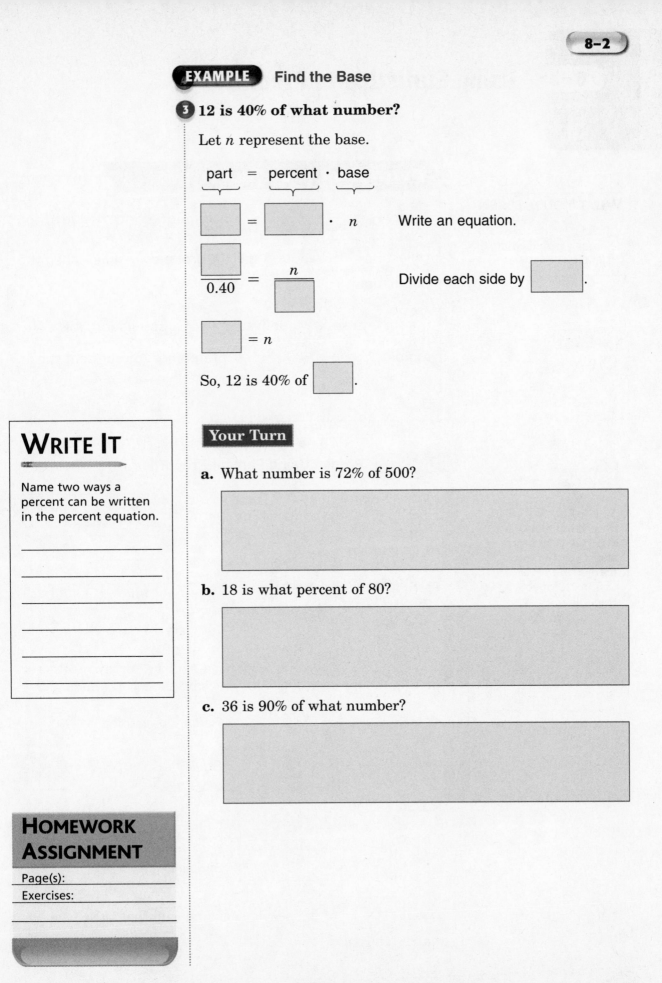

☐ = ☐ · *n*    Write an equation.

$\dfrac{\boxed{\phantom{0}}}{0.40} = \dfrac{n}{\boxed{\phantom{0}}}$    Divide each side by ☐.

☐ = *n*

So, 12 is 40% of ☐.

**WRITE IT**

Name two ways a percent can be written in the percent equation.

_____

_____

_____

_____

_____

_____

**Your Turn**

**a.** What number is 72% of 500?

**b.** 18 is what percent of 80?

**c.** 36 is 90% of what number?

**HOMEWORK ASSIGNMENT**

Page(s):

Exercises:

_____

_____

# Using Statistics to Predict

**BUILD YOUR VOCABULARY** (Page 185)

**WHAT YOU'LL LEARN**

• Predict actions of a larger group by using a sample.

A **survey** is a [          ] or set of questions designed to collect [          ] about a specific group of people, called the **population**.

If a survey uses a **random sample** of a population, it is a sample chosen [          ] preference to represent the [          ].

**FOLDABLES**

## ORGANIZE IT

Record the main ideas, and give examples about using statistics to predict in the row for Lesson 8–3 of your Foldable.

| | Ch 8 | 8–1 | 8–2 | 8–3 | 8–4 | 8–5 | 8–6 | Other Facts |
|---|---|---|---|---|---|---|---|---|
| Main Ideas | | | | | | | | |
| Examples | | | | | | | | |

**EXAMPLE** Predict Using Percent Proportion

① **PETS** The table shows the results of a survey in which people were asked whether their house pets watch television. Predict how many of the 540 people surveyed said their pets watch TV.

| Does your pet watch television? | |
|---|---|
| **Response** | **Percent** |
| yes | 38% |
| no | 60% |
| don't know | 2% |

You can use the percent proportion and the survey results to predict the number of people who said their pets watch TV.

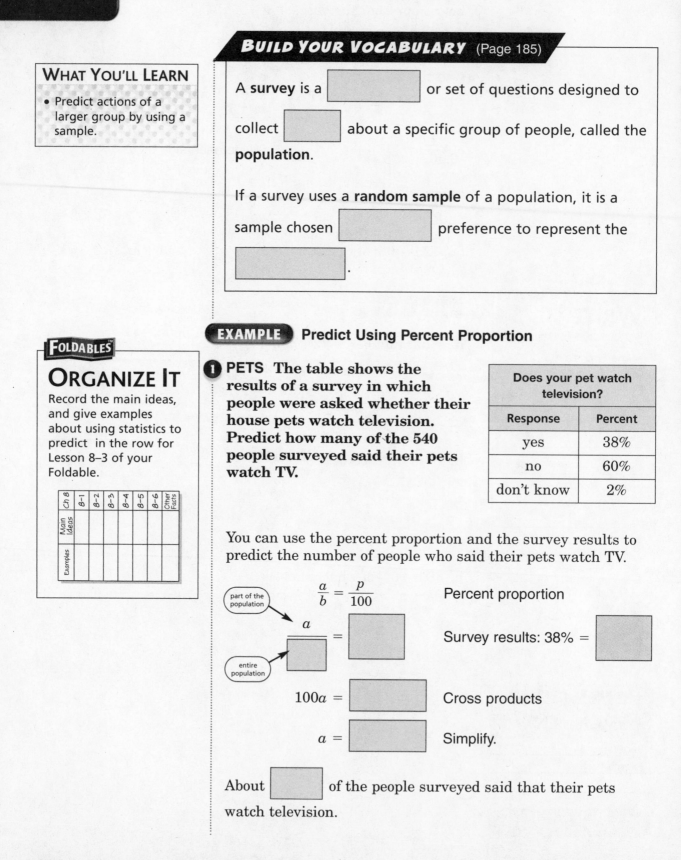

part of the population

$$\frac{a}{b} = \frac{p}{100}$$    Percent proportion

entire population

$$\frac{a}{\boxed{\phantom{xx}}} = \boxed{\phantom{xx}}$$    Survey results: 38% = [          ]

$$100a = \boxed{\phantom{xx}}$$    Cross products

$$a = \boxed{\phantom{xx}}$$    Simplify.

About [          ] of the people surveyed said that their pets watch television.

**Your Turn** In a survey of middle school students, 32% responded that playing video games was their favorite after-school activity. Predict how many of the 260 students surveyed said that playing video games was their favorite after-school activity.

---

**EXAMPLE**   Predict Using Percent Equation

**2** SUMMER JOBS **According to one survey, 25% of high school students reported they would not get a summer job. Predict how many of the 948 students at Mohawk High School will not get a summer job.**

You need to predict how many of the [ ] students will not get a summer job.

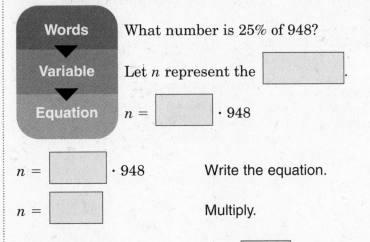

| Words | What number is 25% of 948? |
| Variable | Let $n$ represent the [ ]. |
| Equation | $n =$ [ ] $\cdot$ 948 |

$n =$ [ ] $\cdot$ 948       Write the equation.

$n =$ [ ]        Multiply.

So, you could predict that about [ ] of the students at Mohawk High School will not get a summer job.

**Your Turn** According to one survey, 31% of adults consider spring to be their favorite season of the year. Predict how many of the 525 employees of a large corporation would respond that spring is their favorite season of the year.

# Percent of Change

## WHAT YOU'LL LEARN

- Find the percent of increase or decrease.

## BUILD YOUR VOCABULARY (page 184)

If the [          ] quantity is [          ], the percent of change is called the **percent of increase**.

If the [          ] quantity is [          ], the percent of change is called the **percent of decrease**.

**EXAMPLE** Find Percent of Increase

## KEY CONCEPT

A **percent of change** is a ratio that compares the change in quantity to the original amount.

1. **SHOPPING** Last year a sweater sold for $56. This year the same sweater sells for $60. Find the percent of change in the cost of the sweater. Round to the nearest whole percent if necessary.

Since the new price is [          ] than the original price,

this is a percent of [          ]. The amount of increase is

$60 -$ [          ] or [          ].

$$\text{percent of increase} = \frac{\text{amount of increase}}{\boxed{\phantom{xxxxxxxxxxx}}}$$

$$= \frac{\boxed{\phantom{xx}}}{56} \quad \text{Substitution}$$

$$= \boxed{\phantom{xx}} \quad \text{Simplify.}$$

$$= \boxed{\phantom{xx}} \quad \text{Write as a } \boxed{\phantom{xxxxx}}.$$

The percent of [          ] in the price of the sweater is

about [          ].

**Your Turn** Last year a DVD sold for $20. This year the same DVD sells for $24. Find the percent of change in the cost of the DVD. Round to the nearest whole percent if necessary.

<br>

**EXAMPLE** Find Percent of Decrease

**2 ATTENDANCE** On the first day of school this year, 435 students reported to Howard Middle School. Last year on the first day, 460 students attended. Find the percent of change for the first day attendance. Round to the nearest whole percent if necessary.

Since the new enrollment figure is [____] than the figure for [____] year, this is a percent of [____]. The amount of decrease is [____] − 435 or [____] students.

$$\text{percent of decrease} = \frac{\boxed{\phantom{xxxxxx}}}{\text{original amount}}$$

$$= \frac{25}{\boxed{\phantom{x}}} \qquad \text{Substitution}$$

$$= \boxed{\phantom{x}} \qquad \text{Simplify.}$$

$$= \boxed{\phantom{x}} \qquad \text{Write } \boxed{\phantom{x}} \text{ as a percent.}$$

The percent of [____] in the enrollment is about [____].

**Your Turn** At the beginning of the summer season, the local zoo reported having 385 animals in its care. At the beginning of last year's summer season the zoo had reported 400 animals. Find the percent of change in the number of animals at the zoo. Round to the nearest whole percent if necessary.

**HOMEWORK ASSIGNMENT**
Page(s):
Exercises:

## 8–5 Sales Tax and Discount

**WHAT YOU'LL LEARN**

- Solve problems involving sales tax and discount.

BUILD YOUR VOCABULARY (page 185)

**BUILD YOUR VOCABULARY** (page 185)

**Sales tax** is an [ ] amount of money charged by local, state and federal [ ] on items that people [ ].

**Discount** is the amount by which the regular [ ] of an item is [ ].

**FOLDABLES**

## ORGANIZE IT

Record the main ideas, and give examples about sales tax and discount in the row for Lesson 8–5 of your Foldable.

| | Ch 8 | 8-1 | 8-2 | 8-3 | 8-4 | 8-5 | 8-6 | Other Facts |
|---|---|---|---|---|---|---|---|---|
| Main Ideas | | | | | | | | |
| Examples | | | | | | | | |

**EXAMPLE** Find the Total Cost

**1** **GOLF** A set of golf balls sells for $20 and the sales tax is 5.75%. What is the total cost of the set?

First, find the [ ] tax.

5.75% of $20 = [ ] · 20

= [ ]     The sales tax is [ ].

Next, add the sales tax to the regular price.

[ ] + 20 = [ ]

The [ ] cost of the set of golf balls is [ ].

**Your Turn** A set of three paperback books sells for $35 and the sales tax is 7%. What is the total cost of the set?

**194** *Mathematics: Applications and Concepts, Course 2*

**EXAMPLE** **Find the Sale Price**

2 **Whitney wants to buy a new coat that has a regular price of $185. This weekend, the coat is on sale at a 33% discount. What is the sale price of the coat?**

**Method 1**

First, find the amount of the [          ] $d$.

part = percent · base

$d$ = 0.33 · 185     Use the [          ] equation.

$d$ = [          ]     The discount is [          ]

So, the sale price is $185 − [          ] or [          ].

**Method 2**

First, subtract the [          ] of discount from 100%.

100% − [          ] = [          ]

So, the sale price is [          ] of the regular price.

$s$ = [          ] · 185     Use the [          ] equation.

$s$ = [          ]     The sale price is $123.95

So, the sale price of the coat is [          ]

**Your Turn** Alex wants to buy a DVD player that has a regular price of $175. This weekend, the DVD player is on sale at a 20% discount. What is the sale price of the DVD player?

**EXAMPLE** Find the Percent of the Discount

**3 WATCHES** A sports watch with an original price of $86 is on sale for $60.20. What is the percent of discount?

First, find the [                ] of the discount.

$86 − $60.20 = [                ]

Next, use the [                ] equation to find the [                ] discount.

**Words**    $25.80 is what percent of [                ]

**Variable**    Let $n$ represent the percent.

**Equation**    $25.80 = n \cdot$ [        ]

$25.80 = n \cdot$ [        ]      Write the equation.

[        ] $= n$        [        ] each side by [        ]

and simplify.

The percent of discount is [        ].

**Your Turn** A rocking chair with an original price of $375 is on sale for $318.75. What is the percent of discount?

# Simple Interest

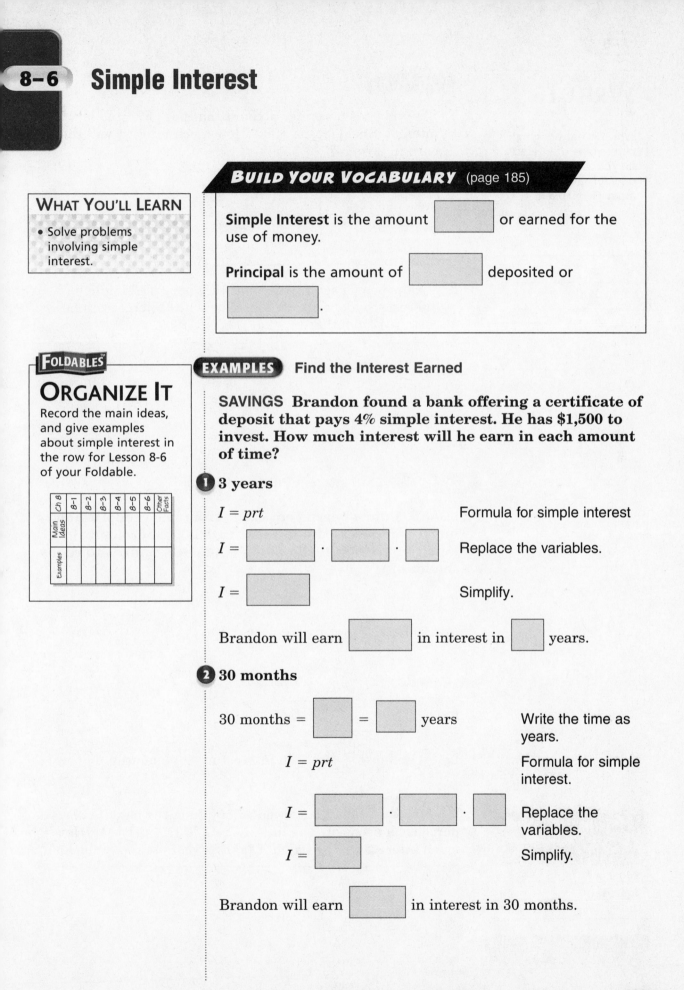

**WHAT YOU'LL LEARN**

• Solve problems involving simple interest.

**FOLDABLES**

## ORGANIZE IT

Record the main ideas, and give examples about simple interest in the row for Lesson 8-6 of your Foldable.

**BUILD YOUR VOCABULARY** (page 185)

**Simple Interest** is the amount [ ] or earned for the use of money.

**Principal** is the amount of [ ] deposited or

[ ] .

**EXAMPLES**   Find the Interest Earned

**SAVINGS**  Brandon found a bank offering a certificate of deposit that pays 4% simple interest. He has $1,500 to invest. How much interest will he earn in each amount of time?

**①** **3 years**

$I = prt$                              Formula for simple interest

$I = $ [ ] · [ ] · [ ]            Replace the variables.

$I = $ [ ]                          Simplify.

Brandon will earn [ ] in interest in [ ] years.

**②** **30 months**

30 months = [ ] = [ ] years       Write the time as years.

$I = prt$                          Formula for simple interest.

$I = $ [ ] · [ ] · [ ]        Replace the variables.

$I = $ [ ]                      Simplify.

Brandon will earn [ ] in interest in 30 months.

# WRITE IT

Which is better: a higher percentage of interest on your credit card or on your savings account? Explain.

_____

_____

_____

_____

## Your Turn

**a.** Cheryl opens a savings account that pays 5% simple interest. She deposits $600. How much interest will she earn in 2 years?

**b.** Micah opens a savings account that pays 4% simple interest. He deposits $2,000. How much interest will he earn in 42 months?

**EXAMPLE** Find Interest Paid on a Loan

**3** LOANS Laura borrowed $2,000 from her credit union to buy a computer. The interest rate is 9% per year. How much interest will she pay if it takes 8 months to repay the loan?

$I = $ ☐          Formula for simple interest

$I = 2,000 \cdot 0.09 \cdot \dfrac{8}{12}$     Replace $p$ with ☐ , $r$ with ☐ , and $t$ with ☐ .

$I = $ ☐          Simplify.

Laura will pay ☐ in interest in ☐ months.

## Your Turn

Juan borrowed $7,500 from the bank to purchase a used car. The interest rate is 15% per year. How much interest will he pay if it takes 2 years to repay the loan?

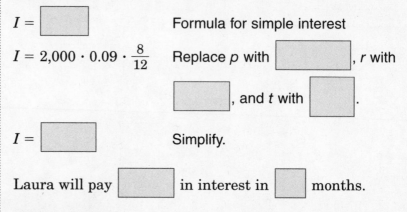

# HOMEWORK ASSIGNMENT

Page(s):

Exercises:

_____

_____

CHAPTER 8

# BRINGING IT ALL TOGETHER

## STUDY GUIDE

| **FOLDABLES** | **VOCABULARY PUZZLEMAKER** | **BUILD YOUR VOCABULARY** |
|---|---|---|
| Use your **Chapter 8 Foldable** to help you study for your chapter test. | To make a crossword puzzle, word search, or jumble puzzle of the vocabulary words in Chapter 8, go to:<br><br>www.glencoe.com/sec/math/t_resources/free/index.php | You can use your completed **Vocabulary Builder** *(pages 184–185)* to help you solve the puzzle. |

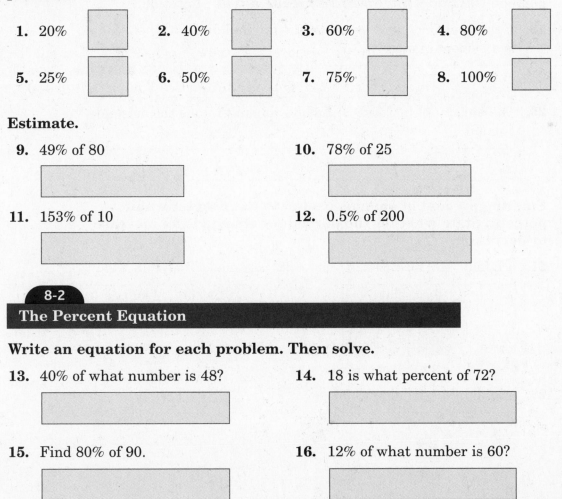

**8-1**

### Percent and Estimation

Write fraction equivalents in simplest form for the following percents.

**1.** 20%  ☐   **2.** 40%  ☐   **3.** 60%  ☐   **4.** 80%  ☐

**5.** 25%  ☐   **6.** 50%  ☐   **7.** 75%  ☐   **8.** 100%  ☐

Estimate.

**9.** 49% of 80

**10.** 78% of 25

**11.** 153% of 10

**12.** 0.5% of 200

**8-2**

### The Percent Equation

Write an equation for each problem. Then solve.

**13.** 40% of what number is 48?

**14.** 18 is what percent of 72?

**15.** Find 80% of 90.

**16.** 12% of what number is 60?

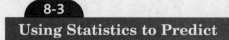

## Using Statistics to Predict

**17. LUNCHES** A survey of seventh graders showed that 44% bring their lunch to school. Predict how many of the 450 seventh graders bring their lunch to school.

**18. ZOO** A survey of visitors at the zoo showed that 28% chose the lion exhibit as their favorite. If 338 people visited the zoo today, predict how many of them would choose the lion exhibit as their favorite.

**8-4**

## Percent of Change

**State whether each sentence is true or false. If false, replace the underlined word to make a true sentence.**

**19.** If the new amount is less than the original amount, then there is a percent of <u>increase</u>.

**20.** The amount of increase is the new amount <u>minus</u> the original amount.

**Find the percent of change. Round to the nearest whole percent. State whether the percent of change is an increase or decrease.**

**21.** original: $48; new $44.25

**22.** original; $157; new $181

**23.** original; $17.48; new $9.98

**8-5**

## Sales Tax and Discount

**Find the total cost or sale price to the nearest cent.**

**24.** $29.99 jeans; 15% discount

**25.** $6.25 lunch; 8.5% sales tax

**Find the percent of discount to the nearest percent.**

**26.** Pen: regular price, $9.95; sale price, $6.95

**27.** Sweatshirt: regular price, $20; sale price, $15.95

**8-6**

## Simple Interest

**Find the interest earned to the nearest cent for each principal, interest rate, and time.**

**28.** $15,000, 9%, 2 years, 4 months

**29.** $250, 3.5%, 6 years

# ARE YOU READY FOR THE CHAPTER TEST?

Visit **msmath2.net** to access your textbook, more examples, self-check quizzes, and practice tests to help you study the concepts in Chapter 8.

Check the one that applies. Suggestions to help you study are given with each item.

☐ **I completed the review of all or most lessons without using my notes or asking for help.**

- You are probably ready for the Chapter Test.

- You may want to take the Chapter 8 Practice Test on page 365 of your textbook as a final check.

☐ **I used my Foldable or Study Notebook to complete the review of all or most lessons.**

- You should complete the Chapter 8 Study Guide and Review on pages 362–364 of your textbook.

- If you are unsure of any concepts or skills, refer back to the specific lesson(s).

- You may also want to take the Chapter 8 Practice Test on page 365 of your textbook.

☐ **I asked for help from someone else to complete the review of all or most lessons.**

- You should review the examples and concepts in your Study Notebook and Chapter 8 Foldable.

- Then complete the Chapter 8 Study Guide and Review on pages 362–364 of your textbook.

- If you are unsure of any concepts or skills, refer back to the specific lesson(s).

- You may also want to take the Chapter 8 Practice Test on page 365 of your textbook.

Student Signature

Parent/Guardian Signature

Teacher Signature

# 9 Probability

 Use the instructions below to make a Foldable to help you organize your notes as you study the chapter. You will see Foldable reminders in the margin of this Interactive Study Notebook to help you in taking notes.

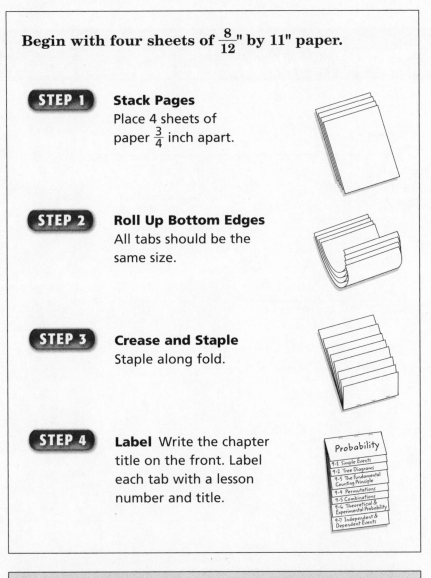

**Begin with four sheets of $\frac{8}{12}$" by 11" paper.**

**STEP 1  Stack Pages**
Place 4 sheets of paper $\frac{3}{4}$ inch apart.

**STEP 2  Roll Up Bottom Edges**
All tabs should be the same size.

**STEP 3  Crease and Staple**
Staple along fold.

**STEP 4  Label** Write the chapter title on the front. Label each tab with a lesson number and title.

Probability
9-1 Simple Events
9-2 Tree Diagrams
9-3 The Fundamental Counting Principle
9-4 Permutations
9-5 Combinations
9-6 Theoretical & Experimental Probability
9-7 Independent & Dependent Events

**NOTE-TAKING TIP:** When taking notes, writing a paragraph that describes the concepts, the computational skills and the graphics will help you to understand the math in a lesson.

**BUILD YOUR VOCABULARY**

This is an alphabetical list of new vocabulary terms you will learn in Chapter 9. As you complete the study notes for the chapter, you will see Build Your Vocabulary reminders to complete each term's definition or description of these pages. Remember to add the textbook page number in the second column for reference when you study.

| Vocabulary Term | Found on Page | Definition | Description or Example |
|---|---|---|---|
| combination | | | |
| complimentary event [KAHM-pluh-MEHN-tuh-ree] | | | |
| compound event | | | |
| dependent event | | | |
| experimental probability [ihk-SPEHR-uh-MEHN-tuhl] | | | |
| factorial [fak-TAWR-ee-uhl] | | | |
| fair game | | | |
| Fundamental Counting Principle | | | |

| Vocabulary Term | Found on Page | Definition | Description or Example |
|---|---|---|---|
| independent event | | | |
| outcome | | | |
| permutation [PUHR-myu-TAY-shuhn] | | | |
| probability [PRAH-buh-BIH-luh-tee] | | | |
| random | | | |
| sample space | | | |
| simple event | | | |
| theoretical probability [thee-uh-REHT-uh-kuhl] | | | |
| tree diagram | | | |

# Simple Events

## WHAT YOU'LL LEARN

• Find the probability of a simple event.

## KEY CONCEPT

**Probability** The probability of an event is a ratio that compares the number of favorable outcomes to the number of possible outcomes.

**FOLDABLES™**

On the tab for Lesson 9-1, takes notes on how to find the probability of simple events. Include examples.

### BUILD YOUR VOCABULARY (page 205)

An **outcome** is any possible [    ].

A **simple event** is one [    ] or a collection of outcomes.

Outcomes occur at **random** if each outcome occurs by [    ].

**EXAMPLE** Find Probability

1. **If the spinner shown is spun once, what is the probability of it landing on an odd number?**

$$P(\text{odd number}) = \frac{\text{odd numbers possible}}{\text{total numbers possible}}$$

$$= \frac{2}{\boxed{\phantom{0}}} \qquad \text{Two numbers are odd: 1 and 3.}$$

$$= \boxed{\phantom{0}} \qquad \text{Simplify.}$$

The probability of spinning an odd number is $\frac{1}{2}$ or [    ].

**Your Turn** What is the probability of rolling a number less than three on a number cube marked with 1, 2, 3, 4, 5, and 6 on its faces?

Two [ ] that are the [ ] ones that can possibly happen are examples of **complementary events**.

**EXAMPLE** Find a Complementary Event

**REVIEW IT**

Explain how to subtract a fraction from 1.

_____

_____

_____

_____

_____

_____

2 **GAMES** A game requires spinning the spinner shown twice. If the sum of the two spins is 6 or greater, the player wins. What is the probability of *not* winning the game?

$P(A) + P(\text{not } A) = 1$

$\dfrac{3}{8} + P(\text{not } A) = 1$      Substitute [ ] for $P(A)$.

$- [\ ] \qquad\qquad - [\ ]$      Subtract $\dfrac{3}{8}$ from each side.
———————————

$P(\text{not } A) = [\ ]$      Simplify.

The probability of not winning the game is [ ].

**Your Turn** A game requires rolling a number cube marked with 1, 2, 3, 4, 5, and 6 on its faces twice. If the sum of the two rolls is five or less, the player wins. What is the probability of winning the game?

**HOMEWORK ASSIGNMENT**

Page(s):

Exercises:

_____

_____

# Tree Diagrams

**BUILD YOUR VOCABULARY** (pages 204–205)

## WHAT YOU'LL LEARN

• Use tree diagrams to count outcomes and find probabilities.

A game in which players of equal skill have an [ ] chance of winning is a **fair game**.

One way of [ ] whether games are fair is by drawing a **tree diagram**.

**Sample space** is the set of all [ ] outcomes.

**FOLDABLES**

## ORGANIZE IT

On the tab for Lesson 9–2, record what you learn about tree diagrams. Explain how to find probability using a tree diagram.

Probability
9-1 Simple Events
9-2 Tree Diagrams
9-3 The Fundamental Counting Principle
9-4 Permutations
9-5 Combinations
9-6 Theoretical & Experimental Probability
9-7 Independent & Dependent Events

**EXAMPLE** Draw a Tree Diagram

❶ **CHILDREN** A family has two children. Draw a tree diagram to show the sample space of the children's genders. Then determine the probability of the family having two girls.

Make a tree diagram. Let B = boy and G = girl.

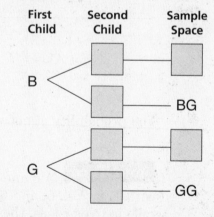

There are [ ] equally-likely outcomes with one showing

two girls. The probability of the family having two girls is [ ].

**Your Turn** A game involves tossing two pennies. Draw a tree diagram to show the sample space of the results in terms of heads and tails. Then determine the probability of tossing one head and one tail.

**EXAMPLE** Find the Number of Outcomes

**2 ICE CREAM** An ice cream sundae at the Ice Cream Shoppe is made from one flavor of ice cream and one topping. For ice cream flavors, you can choose from chocolate, vanilla, and strawberry. For toppings, you can have hot fudge, butterscotch, or marshmallow. Find the number of different sundaes that are possible.

Make a tree diagram. Let C = chocolate, V = vanilla, S = strawberry, H = hot fudge, B = butterscotch, and M = marshmallow.

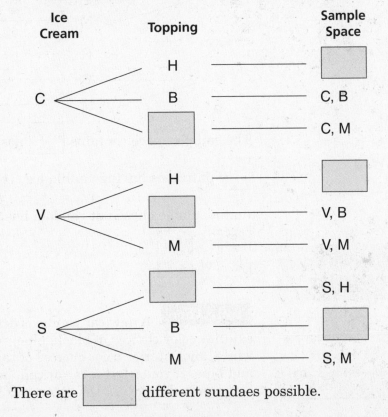

| Ice Cream | Topping | Sample Space |
|---|---|---|
| C | H | |
| | B | C, B |
| | | C, M |
| V | H | |
| | | V, B |
| | M | V, M |
| S | | S, H |
| | B | |
| | M | S, M |

There are [ ] different sundaes possible.

## WRITE IT

In a probability game using two counters A and B, what would the outcome BA mean?

_____

_____

_____

_____

**Your Turn** A new car can be ordered with exterior color choices of black, blue, red, and white, and interior color choices of tan, gray, and blue. Find the number of different cars that are possible.

[                                                                    ]

**EXAMPLE** Find Probability Using Tree Diagrams

**3 ICE CREAM** Refer to Example 2. If you are given a sundae at random from the Ice Cream Shoppe, what is the probability that it has vanilla ice cream?

The tree diagram shows the [                    ]. Find the

probability that the sundae has [          ] ice cream.

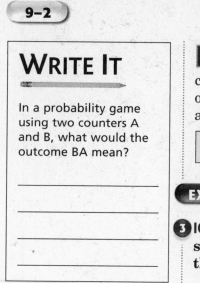

| Ice Cream | Topping | Sample Space |
|---|---|---|
| C | H | C, H |
|  | B | C, B |
|  | M | C, M |
| V | H | V, H |
|  | B | V, B |
|  | M | V, M |
| S | H | S, H |
|  | B | S, B |
|  | M | S, M |

The sample space contains [   ] possible outcomes. There are

[   ] outcomes having vanilla ice cream. So, the probability

that a sundae chosen at random has [      ] ice cream is

[   ] or [   ].

**Your Turn** A new car can be ordered with exterior color choices of black, blue, red, and white, and interior color choices of tan, gray, and blue. If you select a car at random, what is the probability that the exterior color is blue?

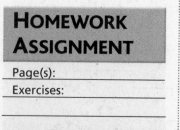

# The Fundamental Counting Principle

## WHAT YOU'LL LEARN

- Use multiplication to count outcomes.

## KEY CONCEPT

**The Fundamental Counting Principle** If event *M* can occur in *m* ways and is followed by event *N* that can occur in *n* ways, then the event *M* followed by *N* can occur in $m \times n$ ways.

**FOLDABLES**

Include this concept in your notes.

**EXAMPLE** Use the Fundamental Counting Principle

① **CLOTHING** The table below shows the shirts, shorts, and shoes in Gerry's wardrobe. How many possible outfits can he choose consisting of one shirt, one pair of shorts, and one pair of shoes?

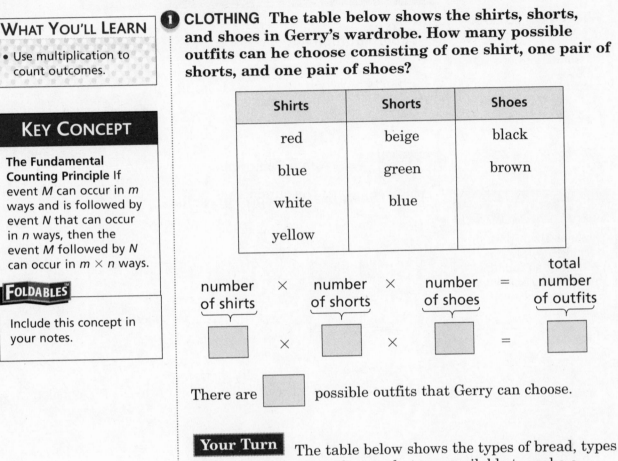

| Shirts | Shorts | Shoes |
|--------|--------|-------|
| red | beige | black |
| blue | green | brown |
| white | blue | |
| yellow | | |

number of shirts  $\times$  number of shorts  $\times$  number of shoes  $=$  total number of outfits

☐  $\times$  ☐  $\times$  ☐  $=$  ☐

There are ☐ possible outfits that Gerry can choose.

**Your Turn** The table below shows the types of bread, types of cheese, and types of meat that are available to make a sandwich. How many possible sandwiches can be made by selecting one type of bread, one type of cheese, and one type of meat?

| Bread | Cheese | Meat |
|-------|--------|------|
| White | American | Turkey |
| Wheat | Swiss | Ham |
| Rye | Mozzarella | Roast Beef |

## HOMEWORK ASSIGNMENT

Page(s):

Exercises:

# 9–4 Permutations

**WHAT YOU'LL LEARN**

• Find the number of permutations of a set of objects.

**BUILD YOUR VOCABULARY** (page 205)

A **permutation** is an [        ], or listing of objects in which [        ] is important.

**EXAMPLES** Evaluate Factorials

Find the value of each expression.

**KEY CONCEPT**

**Factorial** The expression $n$ factorial ($n!$) is the product of all counting numbers beginning with $n$ and counting backward to 1.

**1** 4!

$4! = $ [    ] · [    ] · [    ] · [    ]      Definition of factorial

$\quad = $ [    ]      Simplify.

**2** 3! · 5!

$3! \cdot 5! = $ [    ] · 2 · 1 · [    ] · [    ] · [    ] · 2 · 1   Definition of factorial

$\quad = $ [    ]      Simplify.

**Your Turn** Find the value of each expression.

**a.** 5!

**b.** 4! · 3!

**EXAMPLE** Find a Permutation

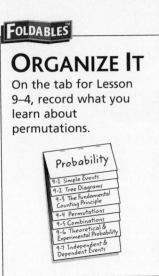
**3 BOWLING** A team of bowlers has five members who bowl one at a time. In how many orders can they bowl?

This is a permutation that can be written as 5!.

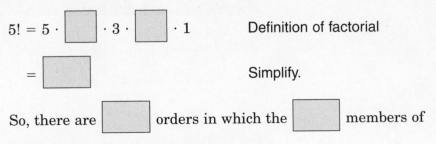

$5! = 5 \cdot \square \cdot 3 \cdot \square \cdot 1$    Definition of factorial

$= \square$    Simplify.

So, there are $\square$ orders in which the $\square$ members of

the bowling team can bowl.

**Your Turn** A relay team has four members who run one at a time. In how many orders can they run?

# 9–5  Combinations

## WHAT YOU'LL LEARN

• Find the number of combinations of a set of objects.

### BUILD YOUR VOCABULARY (page 204)

An arrangement, or listing, of objects in which order is

[_____] is called a **combination**.

---

**FOLDABLES**

## ORGANIZE IT

On the tab for Lesson 9–5, record what you learn about combinations. Be sure to compare and contrast combinations and permutations.

Probability
9-1 Simple Events
9-2 Tree Diagrams
9-3 The Fundamental Counting Principle
9-4 Permutations
9-5 Combinations
9-6 Theoretical & Experimental Probability
9-7 Independent & Dependent Events

---

**EXAMPLE**  Find the Number of Combinations

**1** DECORATING  Ada can select from seven paint colors for her room. She wants to choose two colors. How many different pairs of colors can she choose?

**Method 1**   Make a list.
Number the colors 1 through 7.

| 1, 2 | 1, 5 | 2, 3 | 2, 6 | 3, 5 | 4, 5 | 5, 6 |
| 1, 3 | 1, 6 | 2, 4 | 2, 7 | 3, 6 | 4, 6 | 5, 7 |
| 1, 4 | 1, 7 | 2, 5 | 3, 4 | 3, 7 | 4, 7 | 6, 7 |

There are [____] different pairs of colors.

**Method 2**   Use a permutation.
There are 7 · 6 permutations of two colors chosen from seven. There are 2! ways to arrange the two colors.

$$\frac{7 \cdot 6}{21} = \boxed{\phantom{xx}} = \boxed{\phantom{xx}}$$

There are [____] different pairs of colors Ada can choose.

**Your Turn**   The Brownsville Badgers hockey team has 14 members. Two members of the team are to be selected to be the team's co-captains. How many different pairs of players can be selected to be the co-captains?

[_____]

**REMEMBER IT**

To find a combination you must divide the permutation by the number of ways you can arrange the items.

**EXAMPLES** Identify Permutations and Combinations

Tell whether each situation represents a *permutation* or *combination*. Then solve the problem.

**2** TRACK From an eight-member track team, three members will be selected to represent the team at the state meet. How many ways can these three members be selected?

This is a [ ] because the order of the three

members selected is not important. There are 8 · 7 · 6 ways to choose 3 members. There are 3! ways to arrange 3 people.

$$\frac{8 \cdot 7 \cdot 6}{3!} = \boxed{\phantom{xx}} = \boxed{\phantom{xx}}$$

There are [ ] ways in which the three members can be selected.

**3** In how many ways can you choose the first, second, and third runners in a relay race from eight members of a track team?

This is a [ ] because the order of the runners is important. So, the number of ways the three runners can be

chosen is [ ] · [ ] · [ ] , or [ ] ways.

**Your Turn** Tell whether each situation represents a *permutation* or *combination*. Then solve the problem.

a. There are fifteen members on the PTA for a local middle school. Three of those fifteen will be elected for the offices of president, secretary, and treasurer of the PTA. How many ways can these three positions be filled?

b. In how many ways can you choose a committee of four people from a staff of ten?

**HOMEWORK ASSIGNMENT**

Page(s): _____

Exercises: _____

# Theoretical and Experimental Probability

## WHAT YOU'LL LEARN

- Find and compare experimental and theoretical probabilities.

**Experimental probability** is found using frequencies

obtained in an [ ] or [ ] .

The [ ] probability of an event occurring is called **theoretical probability**.

## ORGANIZE IT

On the tab for Lesson 9–6, take notes about theoretical and experimental probability. Be sure to describe their differences.

Probability

9-1 Simple Events
9-2 Tree Diagrams
9-3 The Fundamental Counting Principle
9-4 Permutations
9-5 Combinations
9-6 Theoretical & Experimental Probability
9-7 Independent & Dependent Events

**EXAMPLE** Experimental Probability

1 **A spinner is spun 50 times, and it lands on the color blue 15 times. What is the experimental probability of spinning blue?**

$$P(\text{blue}) = \frac{\text{number of times } \boxed{\phantom{xx}} \text{ is spun}}{\text{number of } \boxed{\phantom{xx}} \text{ outcomes}}$$

$$= \frac{\boxed{\phantom{xx}}}{\boxed{\phantom{xx}}} \text{ or } \boxed{\phantom{xx}}$$

The experimental probability of spinning the color blue

is [ ] .

**Your Turn** A marble is pulled from a bag of colored marbles 30 times and 18 of the pulls results in a yellow marble. What is the experimental probability of pulling a yellow marble?

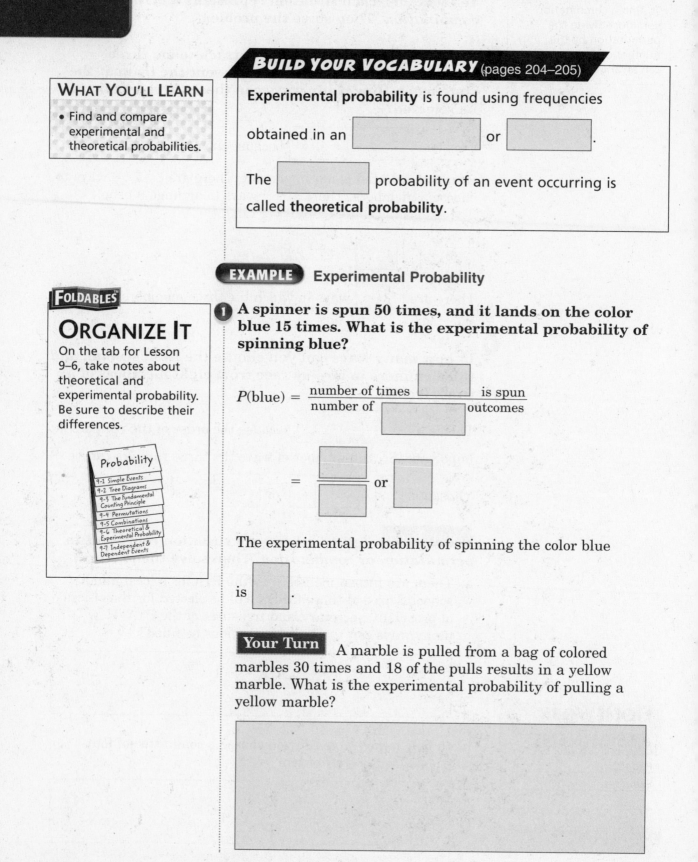

**EXAMPLES** Experimental and Theoretical Probability

The graph shows the results of an experiment in which a number cube is rolled 30 times.

**②** Find the experimental probability of rolling a 5.

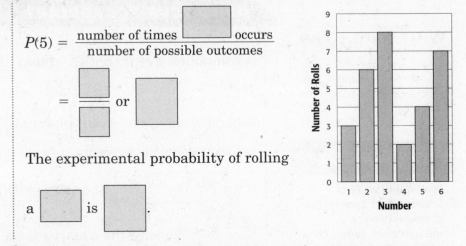

$$P(5) = \frac{\text{number of times } \boxed{\phantom{xx}} \text{ occurs}}{\text{number of possible outcomes}}$$

$$= \frac{\boxed{\phantom{x}}}{\boxed{\phantom{x}}} \text{ or } \boxed{\phantom{x}}$$

The experimental probability of rolling

a $\boxed{\phantom{x}}$ is $\boxed{\phantom{x}}$ .

**③** Compare the experimental probability of rolling a 5 to its theoretical probability.

The theoretical probability of rolling a 5 on a number cube is

$\boxed{\phantom{x}}$ . So, the theoretical probability is close to the

experimental probability of $\boxed{\phantom{x}}$ .

**Your Turn** The graph shows the result of an experiment in which a coin was tossed 150 times.

a. Find the experimental probability of tossing heads for this experiment.

$\boxed{\phantom{xxxxxxxx}}$

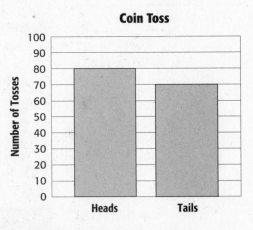

b. Compare the experimental probability of tossing heads to its theoretical probability.

$\boxed{\phantom{xxxxxxxxxxxxxxxxx}}$

**HOMEWORK ASSIGNMENT**

Page(s): _____
Exercises: _____

*Mathematics: Applications and Concepts, Course 2* **217**

**9-7** Independent and Dependent Events

## WHAT YOU'LL LEARN

- Find the probability of independent and dependent events.

## KEY CONCEPT

**Probability of Two Independent Events** The probability of two independent events can be found by multiplying the probability of the first event by the probability of the second event.

**FOLDABLES**

On the tab for Lesson 9-7, give an example of finding the probability of two independent events.

**BUILD YOUR VOCABULARY** (pages 204–205)

A **compound event** consists of two or more events.

When choosing one event does not ⬚ choosing a second event, both events are called **independent events**.

If the ⬚ of one event affects the outcome of a ⬚ event, the events are called **dependent events**.

**EXAMPLE** Independent Events

① **LUNCH** For lunch, Jessica may choose from a turkey sandwich, a tuna sandwich, a salad, or a soup. For a drink, she can choose juice, milk, or water. If she chooses a lunch at random, what is the probability that she chooses a sandwich (of either kind) and juice?

$P$(sandwich) = ⬚          $P$(juice) = ⬚

$P$(sandwich and juice) = ⬚ · ⬚ or ⬚

So, the probability that she chooses a sandwich and juice is ⬚.

**Your Turn** Zachary has a blue, a red, a gray, and a white sweatshirt. He also has blue, red, and gray sweatpants. If Zachary randomly pulls a sweatshirt and a pair of sweatpants from his drawer, what is the probability that they will both be blue?

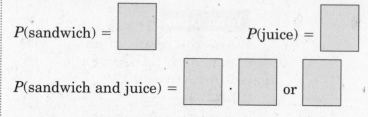

## EXAMPLE  Dependent Events

**KEY CONCEPT**

**Probability of Two Dependent Events** The probability of two dependent events is the probability of the first event times the probability that the second event occurs after the first.

**2** **COMMITTEE SELECTION** Mrs. Tierney will select two students from her class to be on the principal's committee. She places the name of each student in a bag and selects one at a time. The class contains 15 girls and 12 boys. What is the probability she selects a girl's name first, then a boy's name?

There are ▢ students and ▢ are girls. So,

$P$(girl's name first) = ▢ .

There are ▢ students left after one is removed and

▢ are boys. So, $P$(boy's name second) = ▢ or ▢ .

$P$(girl first then boy) = ▢ · ▢ or ▢

So, the probability that Mrs. Tierney will select a girl's name

first and then a boy's name is ▢ , or about ▢ .

**Your Turn**  A box of doughnuts contains 15 glazed doughnuts and 9 jelly doughnuts. Jennifer selects two doughnuts, one at a time. What is the probability that she selects a jelly doughnut first, then a glazed doughnut?

**HOMEWORK ASSIGNMENT**

Page(s):

Exercises:

# CHAPTER 9

# BRINGING IT ALL TOGETHER

## STUDY GUIDE

| **FOLDABLES™** | **VOCABULARY PUZZLEMAKER** | **BUILD YOUR VOCABULARY** |
|---|---|---|
| Use your **Chapter 9 Foldable** to help you study for your chapter test. | To make a crossword puzzle, word search, or jumble puzzle of the vocabulary words in Chapter 9, go to:<br><br>www.glencoe.com/sec/math/t_resources/free/index.php | You can use your completed **Vocabulary Builder** (pages 204–205) to help you solve the puzzle. |

### 9–1

### Simple Events

**Complete each sentence.**

1. A(n) [            ] is a possible result.

2. A complementary event is one of two events that are the only ones that can possibly happen and the sum of those probabilities is [      ] .

**For Questions 3–5, a bag contains 4 green, 6 orange, and 10 purple blocks. Find each probability if you draw one block at random from the bag. Write as a fraction in simplest form.**

3. *P*(green)      4. *P*(orange)      5. *P*(purple)

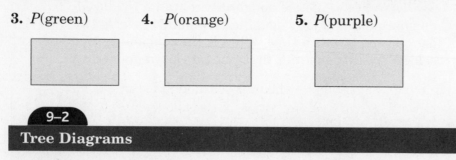

### 9–2

### Tree Diagrams

**Complete each sentence.**

6. A tree diagram resembles a tree because it starts with a base for each event and then branches out to show the possible [            ] of the event.

7. You can use a tree diagram to find the number of possible outcomes by counting the number of entries listed in the tree in the [            ] .

**220**   *Mathematics: Applications and Concepts, Course 2*

**A diner serves tomato, vegetable, and chicken noodle soup and turkey, bologna, cheese, and ham sandwiches.**

8. Make a tree diagram to find the number of possible soup and sandwiches lunches.

**9–3**

## The Fundamental Counting Principle

9. Underline the correct term to complete the sentence: The operation used in the Fundamental Counting Principle is (*addition, multiplication*).

**Use the Fundamental Counting Principle to find the total number of outcomes in each situation.**

10. Tossing a coin and rolling a 6-sided number cube.

11. Making a sandwich using whole wheat or sourdough bread, ham or turkey, and either cheddar, swiss, or provolone cheese.

12. Choosing a marble from a bag containing 10 differently-colored marbles and spinning the spinner at the right.

13. How do you write *five factorial* using symbols?

14. What are the factors of five factorial?

**Find the value of each expression.**

**15.** 6!

**16.** 4! 3!

**17.** 12 · 11 · 10

**18.** A Little League baseball league has 8 teams. In how many ways can the teams finish in first and second place?

---

**9–5**

## Combinations

**Complete each sentence.**

**19.** You can find the number or combinations of objects in a set by

[          ] the number of [          ] of the entire

set by the number of ways each smaller set can be arranged.

**20.** A [          ] is an arrangement or listing in which

order is not [          ].

**21.** The burger shop offers 3 choices of condiments from the following: lettuce, onions, pickles, ketchup, and mustard. How many different combinations of condiments can you have on your burger?

---

**9–6**

## Theoretical and Experimental Probability

**Underline the correct term(s) to complete each sentence.**

**22.** The word experimental means based on (experience, theory).

**23.** Theoretical probability is based on what (you actually try, is expected).

**24.** (Experimental, theoretical) probability can be based on

past performance and can be used to make predictions about future events.

**Sue has 5 different kinds of shoes: sneakers, sandals, boots, moccasins, and heels.**

**25.** If she chooses a pair each day for two weeks, and chooses moccasins 8 times, what is the experimental probability that moccasins are chosen?

**26.** Find the theoretical probability of choosing the moccasins.

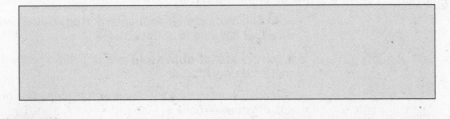

### 9–7
## Independent and Dependent Events

**State whether each sentence is true or false. If false, replace the underlined word to make the sentence true.**

**27.** A <u>compound</u> event consists of more than one single event.

**28.** When the outcome of the first event does not have any effect on the second event it is called a <u>dependent</u> event.

**29.** A yellow and a green cube are rolled. What is the probability that an even number is rolled on the yellow cube and a number less than 3 is rolled on the green cube?

# ARE YOU READY FOR THE CHAPTER TEST?

## Math
### nline

Visit **msmath2.net** to access your textbook, more examples, self-check quizzes, and practice tests to help you study the concepts in Chapter 9.

Check the one that applies. Suggestions to help you study are given with each item.

☐ **I completed the review of all or most lessons without using my notes or asking for help.**

- You are probably ready for the Chapter Test.
- You may want to take the Chapter 9 Practice Test on page 405 of your textbook as a final check.

☐ **I used my Foldable or Study Notebook to complete the review of all or most lessons.**

- You should complete the Chapter 9 Study Guide and Review on pages 402–404 of your textbook.
- If you are unsure of any concepts or skills, refer back to the specific lesson(s).
- You may want to take the Chapter 9 Practice Test on page 405 of your textbook.

☐ **I asked for help from someone else to complete the review of all or most lessons.**

- You should review the examples and concepts in your Study Notebook and Chapter 9 Foldable.
- Then complete the Chapter 9 Study Guide and Review on pages 402–404 of your textbook.
- If you are unsure of any concepts or skills, refer back to the specific lesson(s).
- You may also want to take the Chapter 9 Practice Test on page 405 of your textbook.

Student Signature                    Parent/Guardian Signature

Teacher Signature

# CHAPTER 10

# Geometry

**FOLDABLES** Use the instructions below to make a Foldable to help you organize your notes as you study the chapter. You will see Foldable reminders in the margin of this Interactive Study Notebook to help you in taking notes.

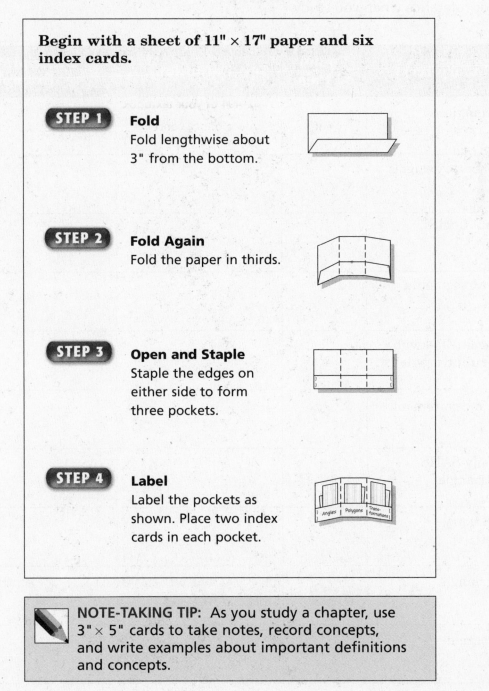

**Begin with a sheet of 11" × 17" paper and six index cards.**

**STEP 1** **Fold**
Fold lengthwise about 3" from the bottom.

**STEP 2** **Fold Again**
Fold the paper in thirds.

**STEP 3** **Open and Staple**
Staple the edges on either side to form three pockets.

**STEP 4** **Label**
Label the pockets as shown. Place two index cards in each pocket.

NOTE-TAKING TIP: As you study a chapter, use 3" × 5" cards to take notes, record concepts, and write examples about important definitions and concepts.

*Chapter 10*

CHAPTER

# 10

## BUILD YOUR VOCABULARY

This is an alphabetical list of new vocabulary terms you will learn in Chapter 10. As you complete the study notes for the chapter, you will see Build Your Vocabulary reminders to complete each term's definition or description on these pages. Remember to add the textbook page number in the second column for reference when you study.

| Vocabulary Term | Found on Page | Definition | Description or Example |
|---|---|---|---|
| acute triangle | 228 | is a angle with Less than 90° Dergrees. | ↗ |
| complementary angles | | | |
| congruent angles | | | |
| congruent segments | | | |
| equilateral [EH-kwuh-LA-tuh-rull] triangle | | | |
| indirect measurement | | | |
| isosceles [y-SAHS-LEEZ] triangle | | | |
| line symmetry | | | |
| obtuse triangle | 228 | an angle more than 90° d regrees. | ↗ |
| parallelogram | | | |

| Vocabulary Term | Found on Page | Definition | Description or Example |
|---|---|---|---|
| quadrilateral [KWAH-druh-LA-tuh-ruhl] | | | |
| reflection | | | |
| rhombus [RAHM-buhs] | | | |
| scalene [SKAY-LEEN] triangle | | | |
| similar figures | | | |
| straight angle | 228 | an angle that is 180° degrees. with a dot on middle | ← • → |
| supplementary angles | | | |
| tessellation | | | |
| translation | | | |
| trapezoid [TRA-puh-ZOYD] | | | |
| vertex | | | |
| vertical angles | | | |

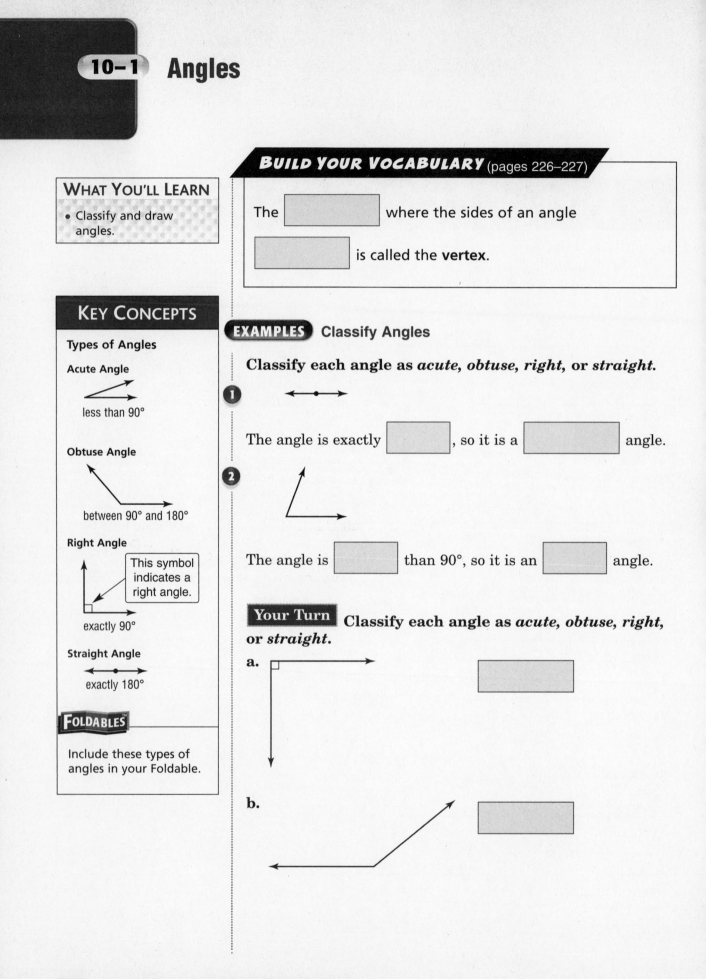

**WHAT YOU'LL LEARN**

• Classify and draw angles.

**KEY CONCEPTS**

**Types of Angles**

**Acute Angle**

less than 90°

**Obtuse Angle**

between 90° and 180°

**Right Angle**

This symbol indicates a right angle.

exactly 90°

**Straight Angle**

exactly 180°

**FOLDABLES**™

Include these types of angles in your Foldable.

**BUILD YOUR VOCABULARY** (pages 226–227)

The [          ] where the sides of an angle

[          ] is called the **vertex**.

**EXAMPLES** Classify Angles

Classify each angle as *acute, obtuse, right,* or *straight.*

**1**

The angle is exactly [          ], so it is a [          ] angle.

**2**

The angle is [          ] than 90°, so it is an [          ] angle.

**Your Turn** Classify each angle as *acute, obtuse, right,* or *straight.*

a. [          ]

b. [          ]

**EXAMPLE**  Draw an Angle

**3** Draw an angle with a measure of 140°. Classify the angle as *acute, obtuse, right,* or *straight.*

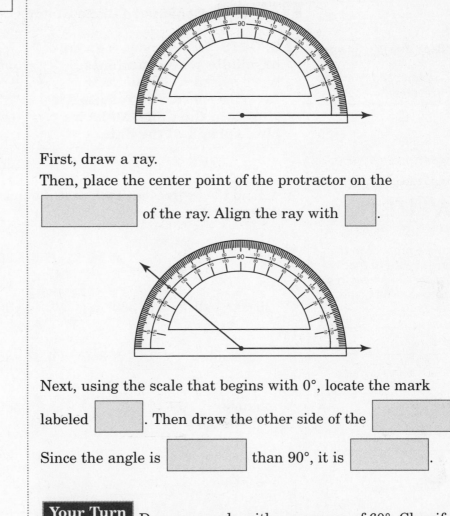

First, draw a ray.

Then, place the center point of the protractor on the

[          ] of the ray. Align the ray with [     ].

Next, using the scale that begins with 0°, locate the mark

labeled [     ]. Then draw the other side of the [          ].

Since the angle is [          ] than 90°, it is [          ].

**Your Turn**  Draw an angle with a measure of 60°. Classify the angle as *acute, obtuse, right,* or *straight.*

**Making Circle Graphs**

**EXAMPLES** Construct a Circle Graph

### WHAT YOU'LL LEARN

• Construct and interpret circle graphs.

**1** SPORTS  In a survey, a group of middle school students were asked to name their favorite sport. The results are shown in the table. Make a circle graph of the data.

| Sport | Percent |
|---|---|
| football | 30% |
| basketball | 25% |
| baseball | 22% |
| tennis | 8% |
| other | 15% |

### WRITE IT

Write a proportion to convert 65% to the number of degrees in a part of a circle graph.

_____

_____

_____

_____

• Find the degrees for each part. Round to the nearest whole degree.

football: [ ] of 360° = 0.30 · 360° or [ ]

basketball: 25% of 360° = [ ] · 360° or [ ]

baseball: [ ] of 360° = 0.22 · 360° or about [ ]

tennis: 8% of 360° = [ ] · 360° or about [ ]

other: [ ] of 360° = 0.15 · 360° or about [ ]

• Use a [ ] to draw a circle with a radius marked as shown. Then use a [ ] to draw the first angle, in this case [ ]. Repeat this step for each section.

- Label each section of the graph with the category and [_____]. Give the graph a [_____].

**Favorite Sport**

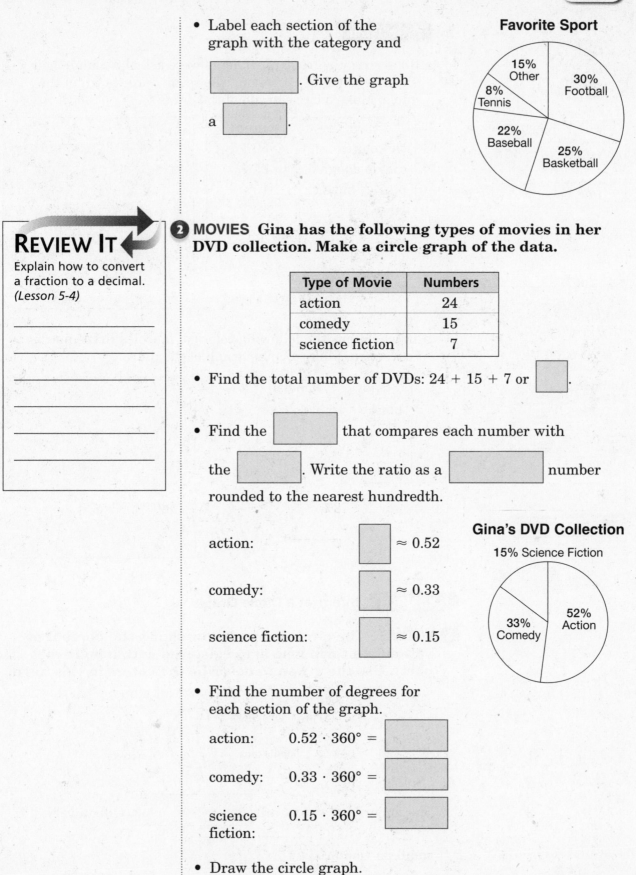

15% Other
8% Tennis
30% Football
22% Baseball
25% Basketball

**REVIEW IT**
Explain how to convert a fraction to a decimal.
*(Lesson 5-4)*

_____
_____
_____
_____
_____
_____

**2 MOVIES** Gina has the following types of movies in her DVD collection. Make a circle graph of the data.

| Type of Movie | Numbers |
|---|---|
| action | 24 |
| comedy | 15 |
| science fiction | 7 |

- Find the total number of DVDs: 24 + 15 + 7 or [___].

- Find the [_____] that compares each number with the [_____]. Write the ratio as a [_____] number rounded to the nearest hundredth.

  action: [___] ≈ 0.52

  comedy: [___] ≈ 0.33

  science fiction: [___] ≈ 0.15

**Gina's DVD Collection**

15% Science Fiction
33% Comedy
52% Action

- Find the number of degrees for each section of the graph.

  action: $0.52 \cdot 360° =$ [_____]

  comedy: $0.33 \cdot 360° =$ [_____]

  science fiction: $0.15 \cdot 360° =$ [_____]

- Draw the circle graph.

**Your Turn**

**a.** In a survey, a group of students were asked to name their favorite flavor of ice cream. The results are shown in the table. Make a circle graph of the data.

| Flavor | Percent |
|---|---|
| chocolate | 30% |
| cookie dough | 25% |
| peanut butter | 15% |
| strawberry | 10% |
| other | 20% |

**b.** Michael has the following colors of marbles in his marble collection. Make a circle graph of the data.

| Color | Number |
|---|---|
| black | 12 |
| green | 9 |
| red | 5 |
| gold | 3 |

**EXAMPLE**   Interpret a Circle Graph

**3** **VOTING**   The circle graph below shows the percent of voters in a town who are registered with a political party. Use the graph to describe the voters in this town.

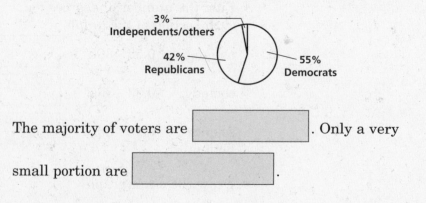

The majority of voters are ⬚⬚⬚⬚⬚⬚. Only a very

small portion are ⬚⬚⬚⬚⬚⬚.

**Your Turn** The circle graph below shows the responses of middle school students to the question "Should teens be allowed to play professional sports?" Use the graph to describe the opinions of the middle school students.

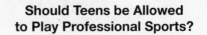

**Should Teens be Allowed to Play Professional Sports?**

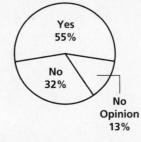

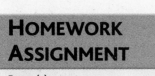

**HOMEWORK ASSIGNMENT**

Page(s): _____

Exercises: _____

_____

_____

# 10–3 Angle Relationships

BUILD YOUR VOCABULARY (pages 226–227)

## WHAT YOU'LL LEARN

- Identify and apply angle relationships.

When two lines [        ], they form two pair of

[        ] angles called **vertical angles**.

Angles with the same [        ] are **congruent angles**.

Two angles are **supplementary** if the sum of their

measures is [      ].

Two angles are **complementary** if the sum of their

measures is [      ].

## FOLDABLES

### ORGANIZE IT

Include examples of supplementary and complementary angles in your Foldable.

Angles | Polygons | Transformations

**EXAMPLES** Classify Angles

Classify each pair of angles as *complementary*, *supplementary*, or *neither*.

1

128°      52°

[        ] + 52° = [        ]

So, the angles are [            ].

2

x
y

∠x and ∠y form a [        ] angle.

So, the angles are [            ].

## REMEMBER IT

When two angles are congruent, the measure of the angles are equal.

**Your Turn** Classify each pair of angles as *complementary, supplementary,* or *neither.*

a.

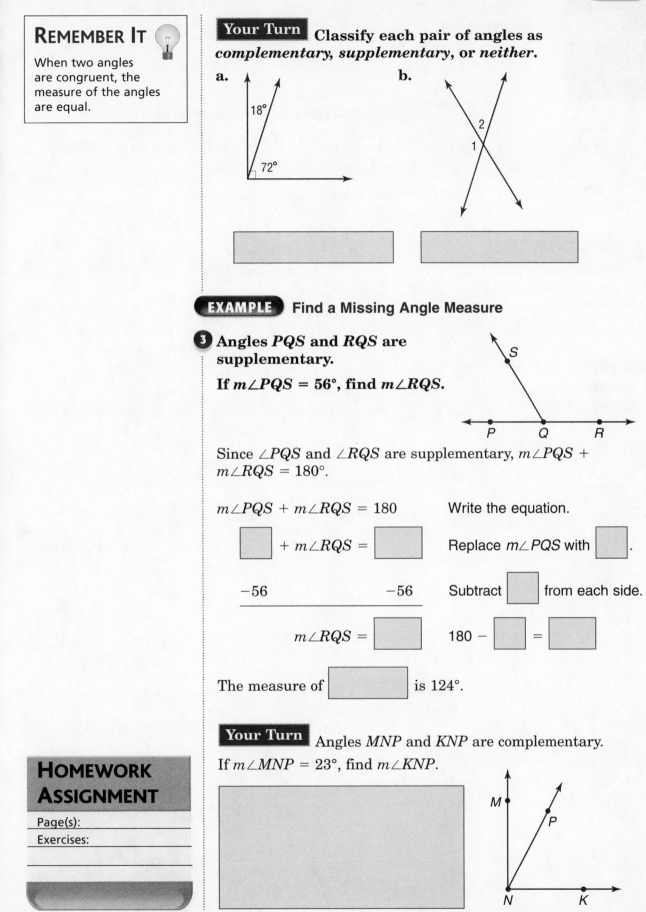

18°

72°

b.

2

1

$\qquad$

$\qquad$

---

**EXAMPLE** Find a Missing Angle Measure

**3** Angles *PQS* and *RQS* are supplementary.

If $m\angle PQS = 56°$, find $m\angle RQS$.

S

P    Q    R

Since $\angle PQS$ and $\angle RQS$ are supplementary, $m\angle PQS + m\angle RQS = 180°$.

$m\angle PQS + m\angle RQS = 180$     Write the equation.

$\boxed{\phantom{xx}} + m\angle RQS = \boxed{\phantom{xx}}$     Replace $m\angle PQS$ with $\boxed{\phantom{xx}}$.

$\underline{-56 \qquad\qquad\qquad -56}$     Subtract $\boxed{\phantom{xx}}$ from each side.

$m\angle RQS = \boxed{\phantom{xx}}$     $180 - \boxed{\phantom{xx}} = \boxed{\phantom{xx}}$

The measure of $\boxed{\phantom{xxx}}$ is 124°.

---

**Your Turn** Angles *MNP* and *KNP* are complementary.

If $m\angle MNP = 23°$, find $m\angle KNP$.

M

P

N    K

---

## HOMEWORK ASSIGNMENT

Page(s):

Exercises:

**Triangles**

**EXAMPLE** Find Angle Measures of Triangles

## KEY CONCEPT

**Angles of a Triangle** The sum of the measures of the angles of a triangle is 180°.

**FOLDABLES**

Record this relationship on a study card. Be sure to include an example.

**1** **PLANES** An airplane has wings that are shaped like triangles. Find the missing measure.

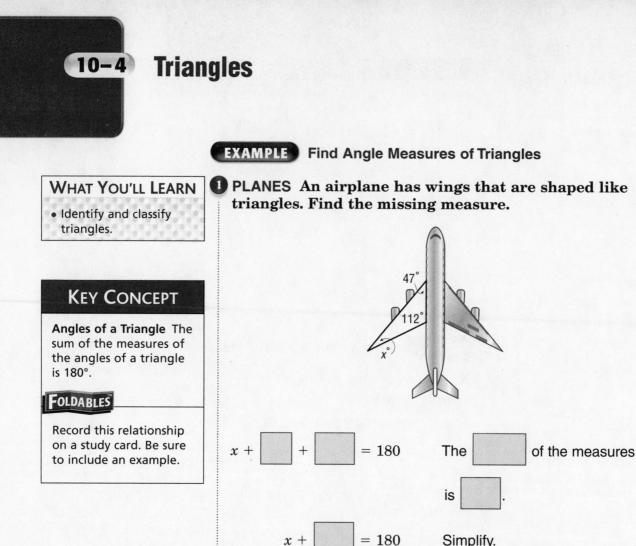

$x + $ ☐ $ + $ ☐ $ = 180$      The ☐ of the measures is ☐.

$x + $ ☐ $ = 180$      Simplify.

☐  ☐      Subtract ☐ from each side.

_____

$x \quad = $ ☐

The missing measure is ☐.

**Your Turn** A piece of fabric is shaped like a triangle. Find the missing measure.

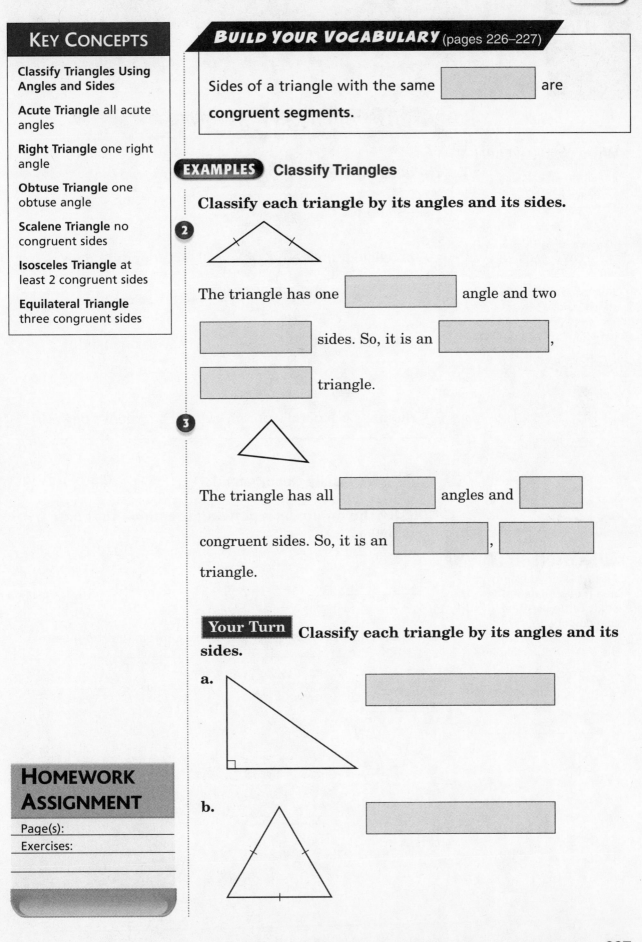

# KEY CONCEPTS

**Classify Triangles Using Angles and Sides**

**Acute Triangle** all acute angles

**Right Triangle** one right angle

**Obtuse Triangle** one obtuse angle

**Scalene Triangle** no congruent sides

**Isosceles Triangle** at least 2 congruent sides

**Equilateral Triangle** three congruent sides

## HOMEWORK ASSIGNMENT

Page(s):
Exercises:

10–4

**BUILD YOUR VOCABULARY** (pages 226–227)

Sides of a triangle with the same [          ] are **congruent segments.**

**EXAMPLES** Classify Triangles

**Classify each triangle by its angles and its sides.**

**2**

The triangle has one [          ] angle and two [          ] sides. So, it is an [          ], [          ] triangle.

**3**

The triangle has all [          ] angles and [          ] congruent sides. So, it is an [          ], [          ] triangle.

**Your Turn** Classify each triangle by its angles and its sides.

**a.**

**b.**

# 10–5 Quadrilaterals

BUILD YOUR VOCABULARY (pages 226–227)

## WHAT YOU'LL LEARN

- Identify and classify quadrilaterals.

A **quadrilateral** is a [          ] figure with [          ] sides and four [          ].

A **parallelogram** is a quadrilateral with opposite sides [          ] and opposite sides [          ].

A **trapezoid** is a [          ] with one pair of [          ] sides.

A **rhombus** is a parallelogram with 4 congruent sides.

---

**EXAMPLES** Classify Quadrilaterals

## FOLDABLES

### ORGANIZE IT

Record what you learn about quadrilaterals on study cards. Illustrate and describe the five types of quadrilaterals discussed in this chapter.

Angles | Polygons | Transformations

Classify the quadrilateral using the name that *best* describes it.

**1**

The quadrilateral has 4 [          ] angles and opposite sides are [          ]. It is a [          ].

**2**

The quadrilateral has [          ] pair of [          ] sides. It is a [          ].

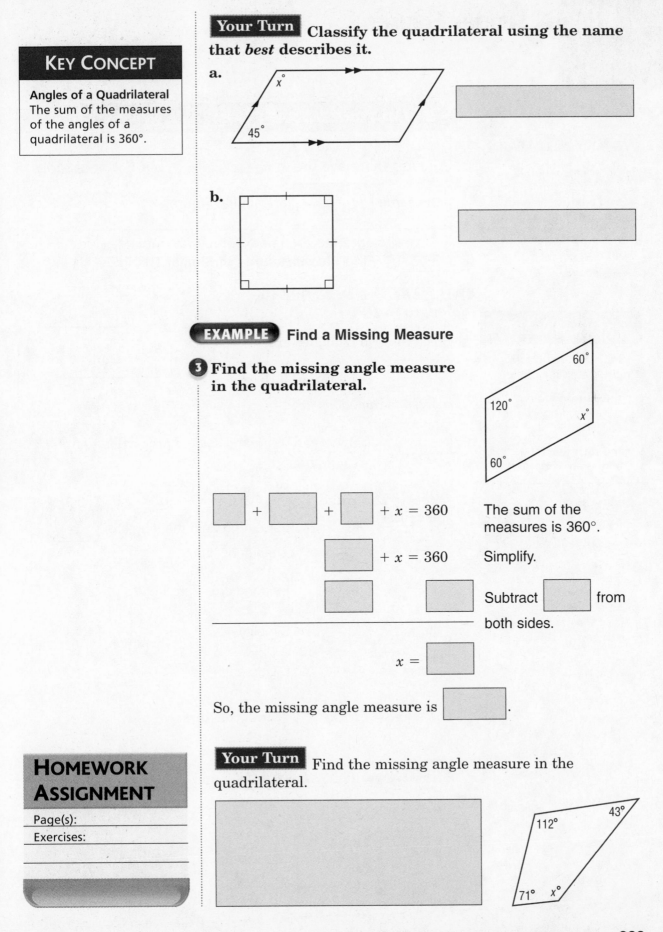

## KEY CONCEPT

**Angles of a Quadrilateral**
The sum of the measures of the angles of a quadrilateral is 360°.

**Your Turn** Classify the quadrilateral using the name that *best* describes it.

a.

x°

45°

b.

**EXAMPLE** Find a Missing Measure

❸ Find the missing angle measure in the quadrilateral.

60°

120°

x°

60°

$\boxed{\phantom{xx}} + \boxed{\phantom{xx}} + \boxed{\phantom{xx}} + x = 360$    The sum of the measures is 360°.

$\boxed{\phantom{xx}} + x = 360$    Simplify.

$\boxed{\phantom{xx}}\qquad\boxed{\phantom{xx}}$    Subtract $\boxed{\phantom{xx}}$ from both sides.

$x = \boxed{\phantom{xx}}$

So, the missing angle measure is $\boxed{\phantom{xx}}$.

## HOMEWORK ASSIGNMENT

Page(s):
Exercises:

**Your Turn** Find the missing angle measure in the quadrilateral.

43°

112°

71° x°

**Similar Figures**

**WHAT YOU'LL LEARN**

• Determine whether figures are similar and find a missing length in a pair of similar figures.

**BUILD YOUR VOCABULARY** (pages 226–227)

Figures that have the same [        ] but not necessarily the same [        ] are **similar figures**.

**EXAMPLE** **Find Side Measures of Similar Triangles**

**KEY CONCEPT**

**Similar Figures** If two figures are similar, then

• the corresponding sides are proportional, and

• the corresponding angles are congruent.

❶ If △ABC ~ △DEF, find the length of $\overline{DF}$.

Since the two triangles are [        ], the ratios of their corresponding sides are

[        ]. So, you can write and solve a proportion to find $\overline{DF}$.

$$\frac{AB}{DE} = \frac{AC}{DF}$$     Write a proportion.

[   ] $= \dfrac{11}{n}$     Let *n* represent the length of [      ]. Then substitute.

$3n =$ [        ]     Find the cross products.

$3n =$ [      ]     Simplify.

$n =$ [      ]     Divide each side by [   ].

The length of $\overline{DF}$ is [      ] centimeters.

**Your Turn** If △JKL ~ △MNO, find the length of $\overline{JL}$.

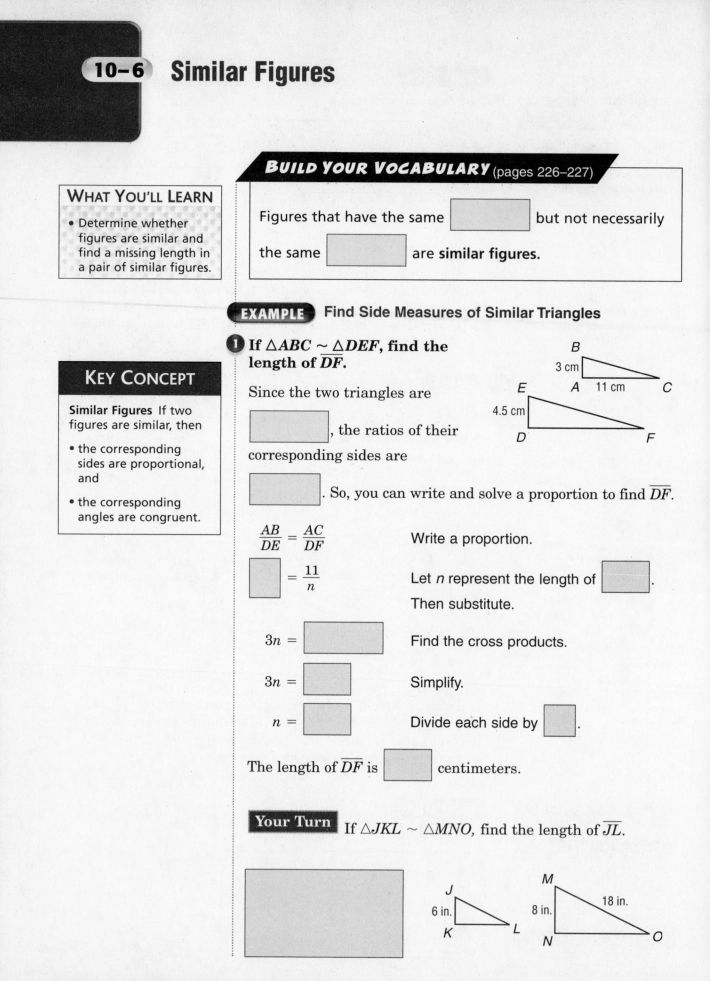

*Mathematics: Applications and Concepts, Course 2*

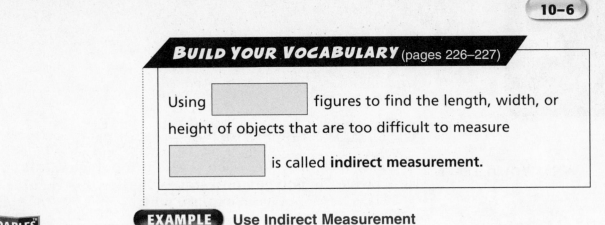

**BUILD YOUR VOCABULARY** (pages 226–227)

Using [ ] figures to find the length, width, or height of objects that are too difficult to measure

[ ] is called **indirect measurement**.

**EXAMPLE** Use Indirect Measurement

② **A rectangular picture window 12-feet long and 6-feet wide needs to be shortened to 9 feet in length to fit a redesigned wall. If the architect wants the new window to be similar to the old window, how wide will the new window be?**

To find the width of the new window, draw a picture and write a proportion.

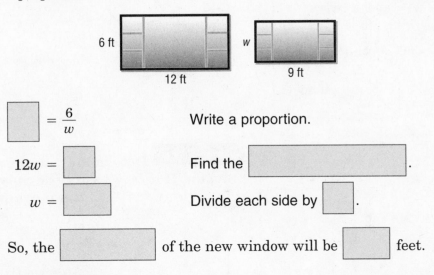

6 ft        12 ft        $w$        9 ft

$$\boxed{\phantom{xx}} = \frac{6}{w}$$     Write a proportion.

$$12w = \boxed{\phantom{xx}}$$     Find the [ ].

$$w = \boxed{\phantom{xx}}$$     Divide each side by [ ].

So, the [ ] of the new window will be [ ] feet.

**Your Turn** Tom has a rectangular garden that has a length of 12 feet and a width of 8 feet. He wishes to start a second garden that is similar to the first and will have a width of 6 feet. Find the length of the new garden.

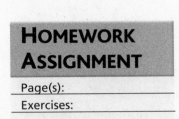

# Polygons and Tessellations

## WHAT YOU'LL LEARN

- Classify polygons and determine which polygons can form a tessellation.

## ORGANIZE IT

Use your study cards to record what you learn about polygons and tessellations. Explain how a tessellation can be made with several kinds of polygons.

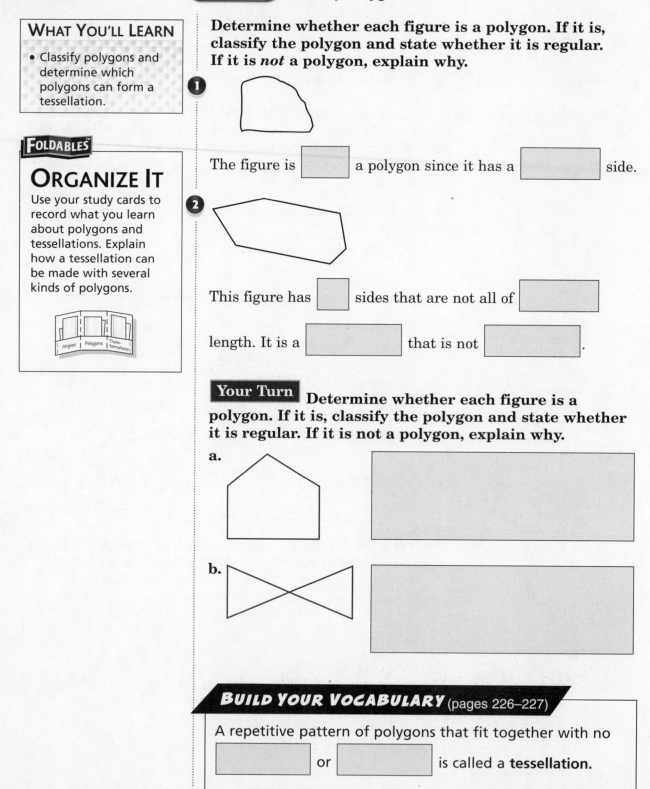

**EXAMPLES** Classify Polygons

Determine whether each figure is a polygon. If it is, classify the polygon and state whether it is regular. If it is *not* a polygon, explain why.

**1**

The figure is [      ] a polygon since it has a [      ] side.

**2**

This figure has [    ] sides that are not all of [      ]

length. It is a [        ] that is not [        ].

**Your Turn** Determine whether each figure is a polygon. If it is, classify the polygon and state whether it is regular. If it is not a polygon, explain why.

a.

b.

**BUILD YOUR VOCABULARY** (pages 226–227)

A repetitive pattern of polygons that fit together with no

[        ] or [        ] is called a **tessellation**.

**EXAMPLE** Tessellations

**3** PATTERNS Ms. Pena is creating a pattern on her wall. She wants to use triangles with angles 120°, 30°, and 30°. Can Ms. Pena tessellate with these triangles?

The [ ] of the measures of the [ ] where the

[ ] meet must be 360°.

Both [ ] and [ ] divide evenly into [ ].

Therefore, Ms. Pena can arrange the triangles in a way that

the angles where the vertices meet make [ ]. She can

[ ] with these triangles.

You can check if your answer is correct by drawing a

[ ] of [ ] with angles measuring

[ ] , [ ] , and [ ] .

Yes, they can be arranged in a way that the [ ] where

the [ ] meet make [ ].

**HOMEWORK ASSIGNMENT**

Page(s):

Exercises:

**Your Turn** Emily is making a quilt using fabric pieces shaped as equilateral triangles. Can Emily tessellate the quilt with these fabric pieces?

# Translations

**BUILD YOUR VOCABULARY** (pages 226–227)

A **translation** is a transformation where every point of the [          ] figure is moved the same [          ] and in the same [          ].

**EXAMPLE** Graph a Translation

1 **Translate △ABC 5 units left and 1 unit up.**

• Move each vertex of the figure 5 units left and 1 unit up. Label the new vertices A′, B′, and C′.

• Connect the vertices to draw the triangle. The coordinates of the vertices of the new figure are [          ], [          ], and [          ].

**Your Turn** Translate △DEF 3 units left and 2 units down.

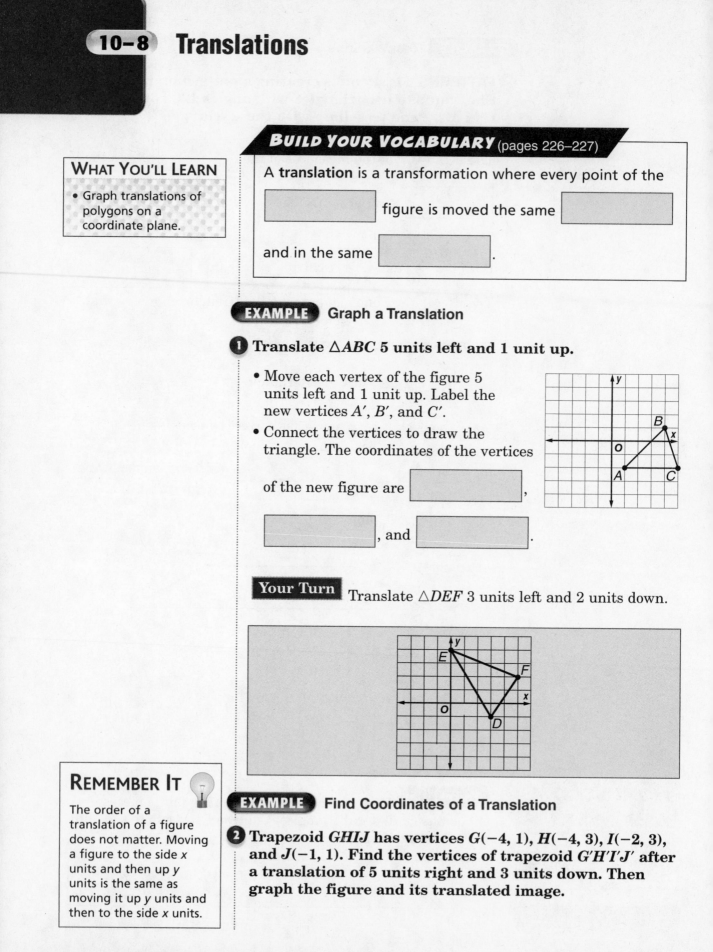

**EXAMPLE** Find Coordinates of a Translation

2 **Trapezoid GHIJ has vertices G(−4, 1), H(−4, 3), I(−2, 3), and J(−1, 1). Find the vertices of trapezoid G′H′I′J′ after a translation of 5 units right and 3 units down. Then graph the figure and its translated image.**

Add ▢ to each *x*-coordinate.

Add ▢ to each *y*-coordinate.

| Vertices of trapezoid *GHIJ* | $(x + 5, y - 3)$ | Vertices of trapezoid *G'H'I'J'* |
|---|---|---|
| $G(-4, 1)$ | | $G'(1, -2)$ |
| $H(-4, 3)$ | $(-4 + 5, 3 - 3)$ | |
| | $(-2 + 5, 3 - 3)$ | |
| $J(-1, 1)$ | | $J'(4, -2)$ |

The coordinates of trapezoid

*G'H'I'J'* are *G'* ▢ , *H'* ▢ ,

*I'* ▢ , and *J'* ▢ .

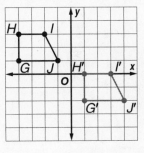

**Your Turn** Triangle *MNO* has vertices $M(-5, -3)$, $N(-7, 0)$, and $O(-2, 3)$. Find the vertices of triangle *M'N'O'* after a translation of 6 units right and 3 units up. Then graph the figure and its translated image.

**HOMEWORK ASSIGNMENT**

Page(s):

Exercises:

## WHAT YOU'LL LEARN

- Identify figures with line symmetry and graph reflections on a coordinate plane.

**BUILD YOUR VOCABULARY** (pages 226–227)

Figures that ☐ exactly when they are folded in ☐ have **line symmetry**.

A type of transformation where a figure is flipped over a line of symmetry is a **reflection**.

**EXAMPLES** Identify Lines of Symmetry

Determine whether each figure has line symmetry. If so, copy the figure and draw all lines of symmetry.

**1** This figure has line ☐.

There are ☐ lines of symmetry.

**2** This figure has line symmetry.

There is ☐ line of symmetry.

**3** This figure ☐ have line symmetry.

**Your Turn** Determine whether each figure has line symmetry. If so, copy the figure and draw all lines of symmetry.

a.

b.

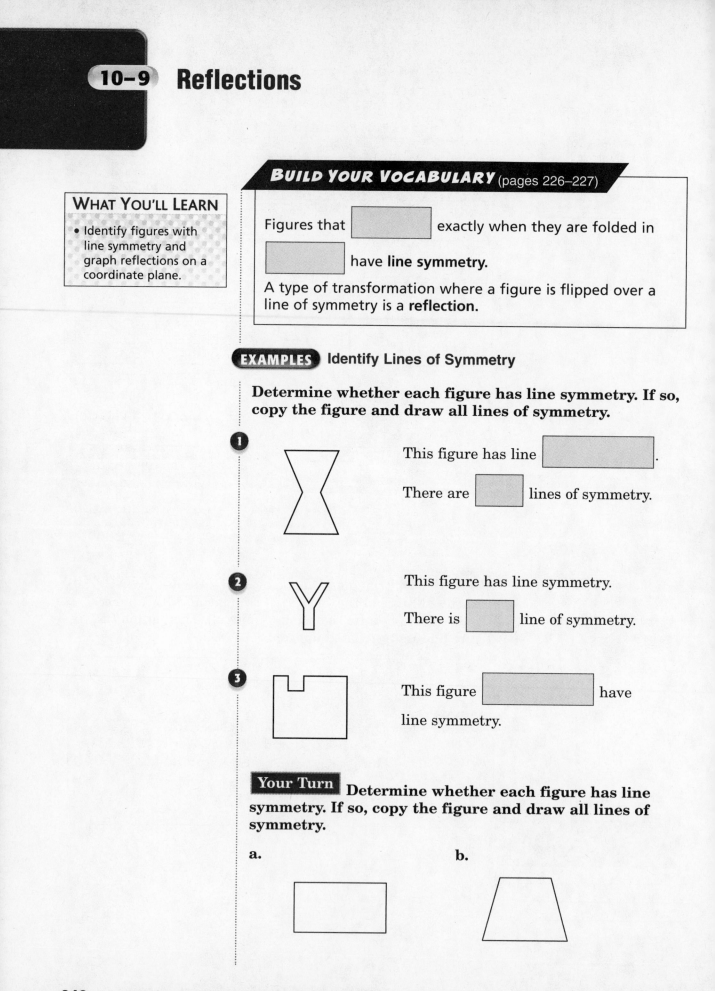

**EXAMPLE** Reflect a Figure Over the *x*-axis

④ **Quadrilateral *QRST* has vertices *Q*(−1, 1), *R*(0, 3), *S*(3, 2), and *T*(4, 0). Find the coordinates of *QRST* after a reflection over the *x*-axis. Then graph the figure and its reflected image.**

| Vertices of Quadrilateral *QRST* | Distance from *x*-axis | Vertices of Quadrilateral *Q'R'S'T'* |
|---|---|---|
| *Q*(−1, 1) |  |  |
| *R*(0, 3) | 3 |  |
| *S*(3, 2) |  |  |
| *T*(4, 0) | 0 |  |

Plot the vertices and connect to form the quadrilateral *QRST*.

The *x*-axis is the line of symmetry. So, the distance from each point on quadrilateral *QRST* to the line of symmetry is the same as the distance from the line of symmetry to quadrilateral *Q'R'S'T'*.

*Q'*

*R'*

*S'*

*T'*

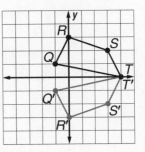

**Your Turn** Quadrilateral *ABCD* has vertices *A*(−3, 2), *B*(−1, 5), *C*(3, 3), and *D*(2, 1). Find the coordinates of ABCD after a reflection over the *x*-axis. Then graph the figure and its reflected image.

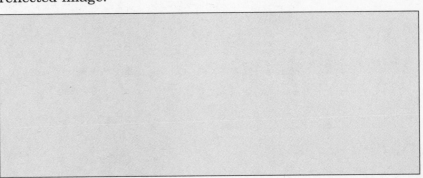

**EXAMPLE** Reflect a Figure over the *y*-axis

**5** Triangle *XYZ* has vertices *X*(1, 2), *Y*(2, 1), and *Z*(1, 2). Find the coordinates of *XYZ* after a reflection over the *y*-axis. Then graph the figure and its reflected image.

| Vertices of △*XYZ* | Distance from *y*-axis | Vertices of △*X'Y'Z'* |
|---|---|---|
| *X*(1, 2) | 1 | |
| *Y*(2, 1) | | |
| *Z*(1, −2) | 1 | |

Plot the _____ and connect to form the triangle _____.

The *y*-axis is the line of symmetry. So, the distance from each point on triangle *XYZ* to the line of symmetry is the same as the distance from the line of symmetry to triangle *X'Y'Z'*.

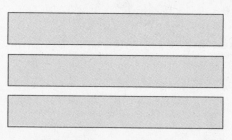

**Your Turn** Triangle *QRS* has vertices *Q*(3, 4), *R*(1, 0), and *S*(6, 2). Find the coordinates of *QRS* after a reflection over the *y*-axis. Then graph the figure and its reflected image.

## STUDY GUIDE

| FOLDABLES™ | VOCABULARY PUZZLEMAKER | BUILD YOUR VOCABULARY |
|---|---|---|
| Use your **Chapter 10 Foldable** to help you study for your chapter test. | To make a crossword puzzle, word search, or jumble puzzle of the vocabulary words in Chapter 10, go to: www.glencoe.com/sec/math/t_resources/free/index.php | You can use your completed **Vocabulary Builder** *(pages 226–227)* to help you solve the puzzle. |

### 10-1
### Angles

**Classify each angle as *acute*, *obtuse*, or *right*.**

1.

2.

3.

### 10-2
### Making Circle Graphs

**Find the number of degrees for each part of the graph in Question 1.**

4. A

5. B

6. C

### 10-3
### Angle Relationships

**Complete each sentence.**

7. The sum of the measures of [ ] angles is 180°.

8. The sum of the measures of [ ] angles is 90°.

9. If $\angle A$ and $\angle B$ are supplementary angles and $m\angle B = 43°$, find $m\angle A$.

**10-4**

## Triangles

Complete the table to help you remember the ways to classify triangles.

| | Type of Triangle | Classified by Angles or Sides | Description |
|---|---|---|---|
| **10.** | acute | angles | |
| **11.** | obtuse | | |
| **12.** | | sides | no congruent sides |
| **13.** | | | 1 right angle |
| **14.** | equilateral | | |

**10-5**

## Quadrilaterals

Find the value of *x* in the quadrilateral.

**15.**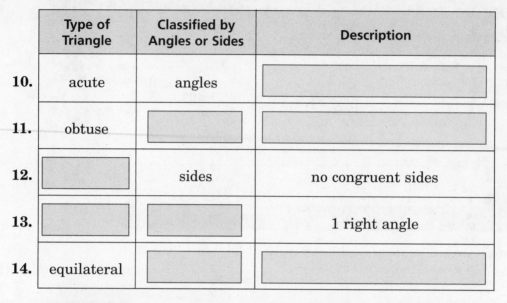

**16.**

**10-6**

## Similar Figures

**17. Find the value of *x*. In each figure, △ABC ~ △DEF.**

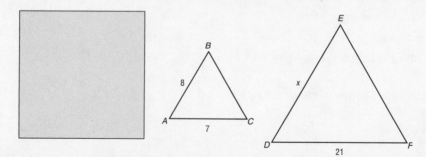

### 10-7
### Polygons and Tessellations

**Underline the correct term to complete each sentence.**

**18.** A polygon can have (two, three) or more straight lines.

**19.** A simple figure is one that (does, does not) have lines that cross each other.

**20.** To find the sum of the angle measures in a regular polygon, draw all the diagonals from one vertex, count the number of (angles, triangles) formed, and multiply by 180°.

### 10-8
### Translations

**State whether the sentence is *true* or *false*. If false, replace the underlined word to make a true sentence.**

**21.** When a figure <u>is turned</u>, every point does not move in the

same direction. ☐

**22.** Triangle *ABC* with vertices *A*(2, 4), *B*(−4, 6), and *C*(1, −5) is translated by (2, −3). What are the coordinates of *B*?

### 10-9
### Reflections

**Underline the correct word(s) to complete each sentence.**

**23.** The image of a reflection is (larger than, the same size as) the original figure.

**24.** If you know the coordinates of a vertex, you can tell how many units it is away from the (*x*-axis, *y*-axis) because it is the same as the *y*-coordinate of the point.

**25.** Triangle *DEF* has vertices *D*(−5, 2), *E*(−4, −2), and *F*(−3, 0). It is reflected over the *y*-axis. What are the coordinates of *D*?

# ARE YOU READY FOR THE CHAPTER TEST?

Visit **msmath2.net** to access your textbook, more examples, self-check quizzes, and practice tests to help you study the concepts in Chapter 10.

Check the one that applies. Suggestions to help you study are given with each item.

☐ **I completed the review of all or most lessons without using my notes or asking for help.**

- You are probably ready for the Chapter Test.

- You may want to take the Chapter 10 Practice Test on page 465 of your textbook as a final check.

☐ **I used my Foldable or Study Notebook to complete the review of all or most lessons.**

- You should complete the Chapter 10 Study Guide and Review on pages 462–464 of your textbook.

- If you are unsure of any concepts or skills, refer to the specific lesson(s).

- You may want to take the Chapter 10 Practice Test on page 465 of your textbook.

☐ **I asked for help from someone else to complete the review of all or most lessons.**

- You should review the examples and concepts in your Study Notebook and Chapter 10 Foldable.

- Then complete the Chapter 10 Study Guide and Review on pages 462–464 of your textbook.

- If you are unsure of any concepts or skills, refer to the specific lesson(s).

- You may also want to take the Chapter 10 Practice Test on page 465 of your textbook.

Student Signature

Parent/Guardian Signature

Teacher Signature

# Geometry: Measuring Two-Dimensional Figures

**FOLDABLES**™ Use the instructions below to make a Foldable to help you organize your notes as you study the chapter. You will see Foldable reminders in the margin of this Interactive Study Notebook to help you in taking notes.

**Begin with a piece of 11" × 17" paper.**

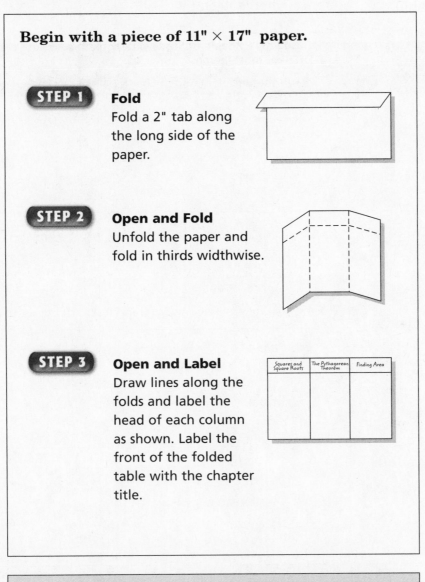

**STEP 1** **Fold**
Fold a 2" tab along the long side of the paper.

**STEP 2** **Open and Fold**
Unfold the paper and fold in thirds widthwise.

**STEP 3** **Open and Label**
Draw lines along the folds and label the head of each column as shown. Label the front of the folded table with the chapter title.

Squares and Square Roots | The Pythagorean Theorem | Finding Area

**NOTE-TAKING TIP:** When you take notes, it is helpful to write key vocabulary words, definitions, concepts, or procedures as clearly and concisely as possible.

This is an alphabetical list of new vocabulary terms you will learn in Chapter 11. As you complete the study notes for the chapter, you will see Build Your Vocabulary reminders to complete each term's definition or description on these pages. Remember to add the textbook page number in the second column for reference when you study.

| Vocabulary Term | Found on Page | Definition | Description or Example |
|---|---|---|---|
| base | | | |
| complex figure | | | |
| height | | | |
| hypotenuse [heye-PAH-tuhn-OOS] | | | |
| irrational number | | | |

| Vocabulary Term | Found on Page | Definition | Description or Example |
|---|---|---|---|
| leg | | | |
| perfect square | | | |
| Pythagorean Theorem [puh-THAG-uh-REE-uhn] | | | |
| radical sign | | | |
| square | | | |
| square roots | | | |

# 11-1 Squares and Square Roots

## WHAT YOU'LL LEARN

- Find squares of numbers and square roots of perfect squares.

**BUILD YOUR VOCABULARY** (page 255)

The [        ] of a number and [        ] is the **square** of the number.

**Perfect squares** are squares of [        ] numbers.

The [        ] multiplied to form perfect squares are called **square roots**.

A **radical sign**, $\sqrt{\phantom{x}}$ , is the symbol used to indicate the positive [        ] of a number.

**EXAMPLES** Find Squares of Numbers

**1** Find the square of 5.

[        ] [        ] = 25

25 units | 5 units

5 units

**2** Find the square of 19.

[        ] $x^2$ ENTER [        ]

**FOLDABLES**

## ORGANIZE IT

In the Squares and Square Roots column of your Foldable, explain in words and symbols how you find squares of numbers and square roots of perfect squares.

| Squares and Square Roots | The Pythagorean Theorem | Finding Area |
|---|---|---|
|  |  |  |

**Your Turn** Find the square of each number.

a. 7

b. 21

**EXAMPLE**  Find a Square to Solve a Problem

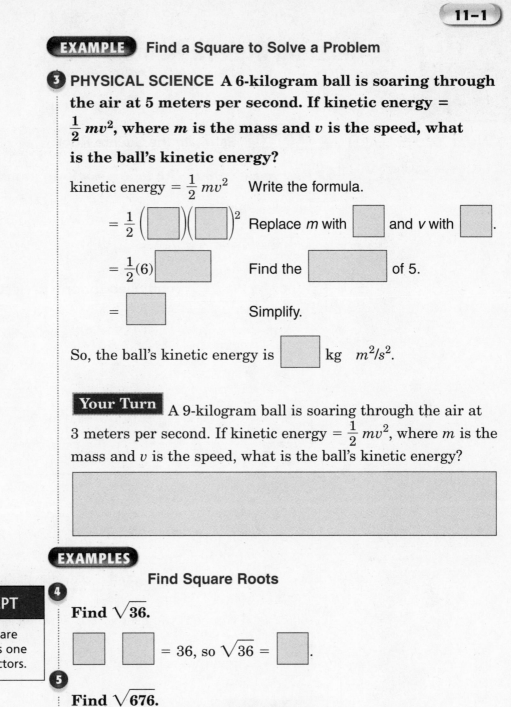

**3** PHYSICAL SCIENCE A 6-kilogram ball is soaring through the air at 5 meters per second. If kinetic energy = $\frac{1}{2}mv^2$, where $m$ is the mass and $v$ is the speed, what is the ball's kinetic energy?

kinetic energy = $\frac{1}{2}mv^2$   Write the formula.

$= \frac{1}{2}\left(\boxed{\phantom{x}}\right)\left(\boxed{\phantom{x}}\right)^2$   Replace $m$ with $\boxed{\phantom{x}}$ and $v$ with $\boxed{\phantom{x}}$.

$= \frac{1}{2}(6)\boxed{\phantom{xx}}$   Find the $\boxed{\phantom{xxx}}$ of 5.

$= \boxed{\phantom{x}}$   Simplify.

So, the ball's kinetic energy is $\boxed{\phantom{x}}$ kg $m^2/s^2$.

**Your Turn** A 9-kilogram ball is soaring through the air at 3 meters per second. If kinetic energy = $\frac{1}{2}mv^2$, where $m$ is the mass and $v$ is the speed, what is the ball's kinetic energy?

**EXAMPLES**

## KEY CONCEPT

**Square Root** A square root of a number is one of its two equal factors.

**Find Square Roots**

**4** Find $\sqrt{36}$.

$\boxed{\phantom{x}}\ \boxed{\phantom{x}} = 36$, so $\sqrt{36} = \boxed{\phantom{x}}$.

**5** Find $\sqrt{676}$.

[2nd] $[\ \sqrt{\phantom{x}}\ ]$ $\boxed{\phantom{xx}}$ [ENTER] $\boxed{\phantom{xx}}$

So, $\sqrt{676} = \boxed{\phantom{x}}$.

## HOMEWORK ASSIGNMENT

Page(s):

Exercises:

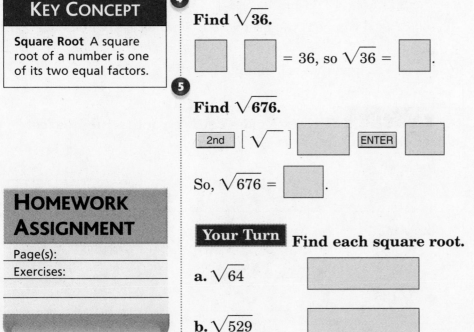

**Your Turn** Find each square root.

**a.** $\sqrt{64}$   $\boxed{\phantom{xxxxxx}}$

**b.** $\sqrt{529}$   $\boxed{\phantom{xxxxxx}}$

**Estimating Square Roots**

**EXAMPLE** Estimate the Square Root

**1** Estimate $\sqrt{96}$ to the nearest whole number.

List some perfect squares.

1, 4, 9, 16, 25, 36, 49, 64, 81, 100 . . .

[box]

$81 \quad < \quad 96 \quad < \quad 100$    96 is between the [box]

squares [box] and [box].

[box] $< \sqrt{96} <$ [box]    Find the $\sqrt{\phantom{x}}$ of each number.

[box] $< \sqrt{96} <$ [box]    [box] $= 9$ and

[box] $= 10$

So, $\sqrt{96}$ is between [box] and [box]. Since 96 is closer

to [box] than 81, the best whole number estimate is

[box]. Verify with a calculator.

**Your Turn** Estimate each square root to the nearest whole number.

a. $\sqrt{41}$

[box]

b. $\sqrt{86}$

[box]

c. $\sqrt{138}$

[box]

## BUILD YOUR VOCABULARY (page 254)

A number that cannot be written as a [ ] is an **irrational number**.

---

**EXAMPLE** Use a Calculator to Estimate

❷ Use a calculator to find the value of $\sqrt{37}$ to the nearest tenth.

[2nd] [ √ ] 37 [ENTER]  [ ]

$\sqrt{37}$  [ ]

Check [ ] = 36 and [ ] = 49. Since [ ] is between 36 and 49, the answer, [ ], is reasonable.

**Your Turn** Use a calculator to find the value of each square root to the nearest tenth.

a. $\sqrt{78}$

b. $\sqrt{96}$

c. $\sqrt{188}$

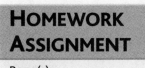

**HOMEWORK ASSIGNMENT**

Page(s):
Exercises:

# The Pythagorean Theorem

**WHAT YOU'LL LEARN**

- Find length using the Pythagorean Theorem.

**BUILD YOUR VOCABULARY** (pages 254–255)

The two sides [        ] to the right [        ] of a right triangle are the **legs**.

The side [        ] the right [        ] of a right triangle is the **hypotenuse**.

The **Pythagorean Theorem** describes the relationship between the length of the [        ] and the lengths of the [        ] of a right triangle.

**EXAMPLE** Find the Length of the Hypotenuse

**KEY CONCEPT**

**Pythagorean Theorem**
In a right triangle, the square of the length of the hypotenuse equals the sum of the squares of the lengths of the legs.

**①** **GYMNASTICS** A gymnastics tumbling floor is in the shape of a square with sides 12 meters long. If a gymnast flips from one corner to the opposite corner, about how far has he flipped?

12 m

*c* m

12 m

To solve, find the length of the hypotenuse *c*.

$$c^2 = a^2 + b^2$$   Pythagorean Theorem

$$c^2 = \boxed{\phantom{xx}} + 12^2$$   Replace *a* with [    ] and *b* with [    ].

$$c^2 = 144 + \boxed{\phantom{xx}}$$   Evaluate [    ].

$$c^2 = \boxed{\phantom{xx}}$$   Add.

$$\sqrt{c^2} = \boxed{\phantom{xx}}$$   Take the [        ] of each side.

$$c \approx \boxed{\phantom{xx}}$$   Simplify.

The gymnast will have flipped about [        ].

**Your Turn** Rose has a rectangular piece of fabric 28 inches long and 16 wide. She wants to decorate the fabric with lace sewn across both diagonals. How much lace will Rose need?

**EXAMPLE** Find the Length of a Leg

**2** Find the missing measure of the triangle at the right.

15 cm    $a$

9 cm

$c^2 = a^2 + b^2$      Pythagorean Theorem

$15^2 = a^2 + 9^2$     Replace $b$ with 9 and $c$ with 15.

☐ $= a^2 +$ ☐     Evaluate ☐ and ☐.

$255 -$ ☐ $= a^2 + 81 -$ ☐     Subtract ☐ from each side.

☐ $= a^2$     Simplify.

$\sqrt{144} = \sqrt{a^2}$     Take the ☐ of each side.

☐ $= a$     Simplify.

The length of the leg is ☐ centimeters.

**Your Turn** Find the missing measure of the triangle. Round to the nearest tenth if necessary.

7 in.    20 in.

$b$ in.

**EXAMPLE** Identify Right Triangles

**Determine whether a triangle with the given side lengths is a right triangle.**

**③ 2.5 cm, 6 cm, 6.5 cm**

$$c^2 = a^2 + b^2$$  Pythagorean Theorem

☐ $\overset{?}{=}$ ☐ + ☐  Replace $a$ with ☐, $b$ with ☐, and $c$ with ☐.

☐ $\overset{?}{=}$ ☐ + ☐  Evaluate squares.

$42.25 = 42.25$ ✓  Simplify.

The triangle ☐ a right triangle.

**④ 5 ft, 6 ft, 8 ft**

$$c^2 = a^2 + b^2$$  Pythagorean Theorem

☐ $\overset{?}{=}$ ☐ + ☐  Replace ☐ with 5, ☐ with 6, and ☐ with 8.

☐ $\overset{?}{=}$ ☐ + ☐  Evaluate ☐.

☐ $\neq 61$  Simplify.

The triangle ☐ a right triangle.

**Your Turn** **Determine whether a triangle with the given side lengths is a right triangle.**

**a.** 5 in., 12 in., 13 in.

**b.** 4.5 cm, 9 cm, 12.5 cm

# Area of Parallelograms

**WHAT YOU'LL LEARN**

• Find the areas of parallelograms.

**BUILD YOUR VOCABULARY** (page 254)

The **base** is a [          ] of a parallelogram.

The **height** is the length of the segment [          ]

to the [          ] with endpoints on [          ] sides.

**EXAMPLE**  Find the Area of a Parallelogram

**KEY CONCEPT**

**Area of a Parallelogram** The area $A$ of a parallelogram equals the product of its base $b$ and height $h$.

**①** Find the area of the parallelogram.

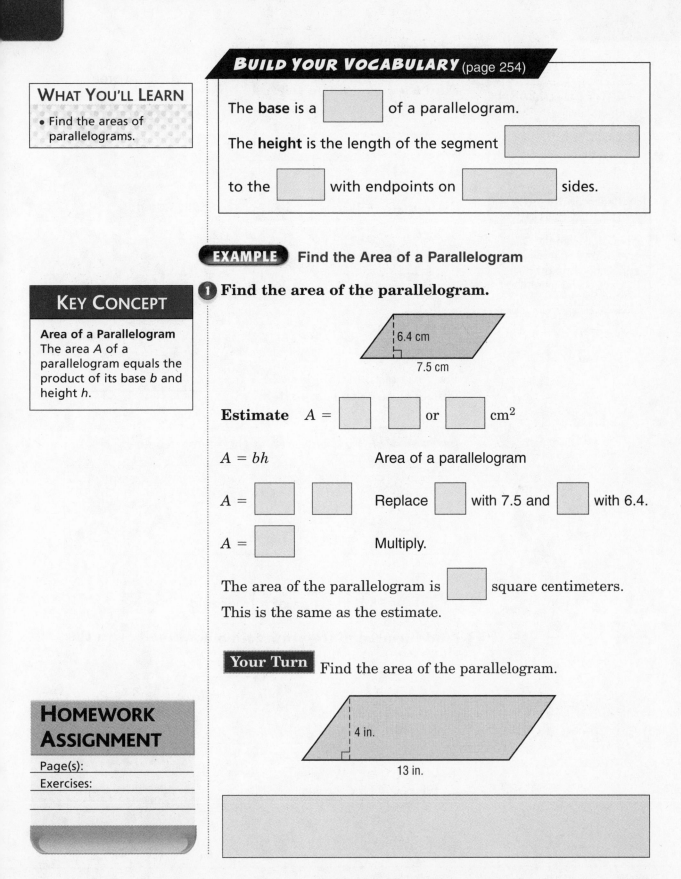

6.4 cm

7.5 cm

**Estimate**   $A =$ [     ] [     ] or [     ] cm²

$A = bh$            Area of a parallelogram

$A =$ [     ] [     ]      Replace [     ] with 7.5 and [     ] with 6.4.

$A =$ [     ]          Multiply.

The area of the parallelogram is [     ] square centimeters. This is the same as the estimate.

**Your Turn**  Find the area of the parallelogram.

4 in.

13 in.

**HOMEWORK ASSIGNMENT**

Page(s): _____
Exercises: _____

# 11-5 Areas of Triangles and Trapezoids

## WHAT YOU'LL LEARN

- Find the areas of triangles and trapezoids.

## KEY CONCEPT

**Area of a Triangle** The area $A$ of a triangle equals half the product of its base $b$ and height $h$.

**EXAMPLE** Find the Area of a Triangle

**1** Find the area of the triangle below.

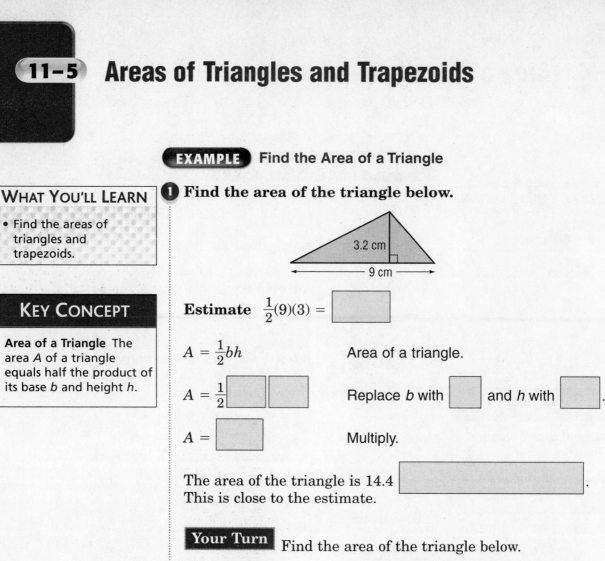

3.2 cm

9 cm

Estimate $\frac{1}{2}(9)(3) = $ ☐

$A = \frac{1}{2}bh$      Area of a triangle.

$A = \frac{1}{2}$ ☐ ☐      Replace $b$ with ☐ and $h$ with ☐ .

$A = $ ☐      Multiply.

The area of the triangle is 14.4 ☐ .
This is close to the estimate.

**Your Turn** Find the area of the triangle below.

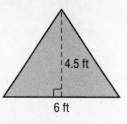

4.5 ft

6 ft

**EXAMPLE** Find the Area of a Trapezoid

**2** Find the area of the trapezoid below.

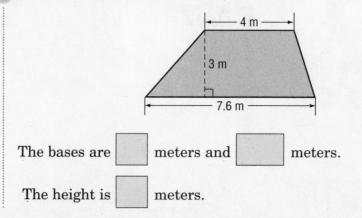

4 m

3 m

7.6 m

The bases are ☐ meters and ☐ meters.

The height is ☐ meters.

## KEY CONCEPT

**Area of a Trapezoid** The area $A$ of a trapezoid equals half the product of the height $h$ and the sum of the bases $b_1$ and $b_2$.

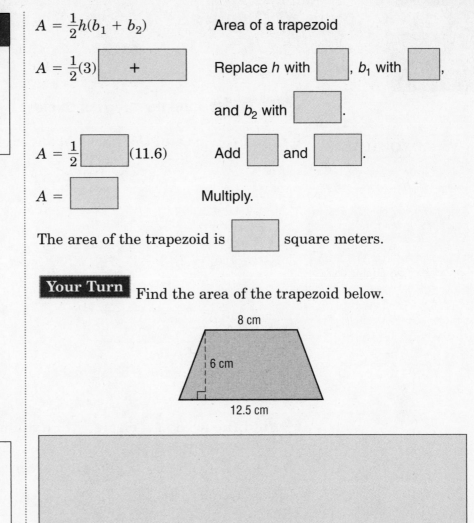

$A = \frac{1}{2}h(b_1 + b_2)$    Area of a trapezoid

$A = \frac{1}{2}(3)$ [  + ]    Replace $h$ with [ ], $b_1$ with [ ],

and $b_2$ with [ ].

$A = \frac{1}{2}$[ ](11.6)    Add [ ] and [ ].

$A = $ [ ]    Multiply.

The area of the trapezoid is [ ] square meters.

**Your Turn** Find the area of the trapezoid below.

8 cm

6 cm

12.5 cm

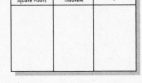
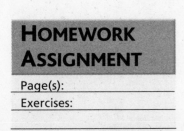

## HOMEWORK ASSIGNMENT

Page(s):

Exercises:

**Area of Circles**

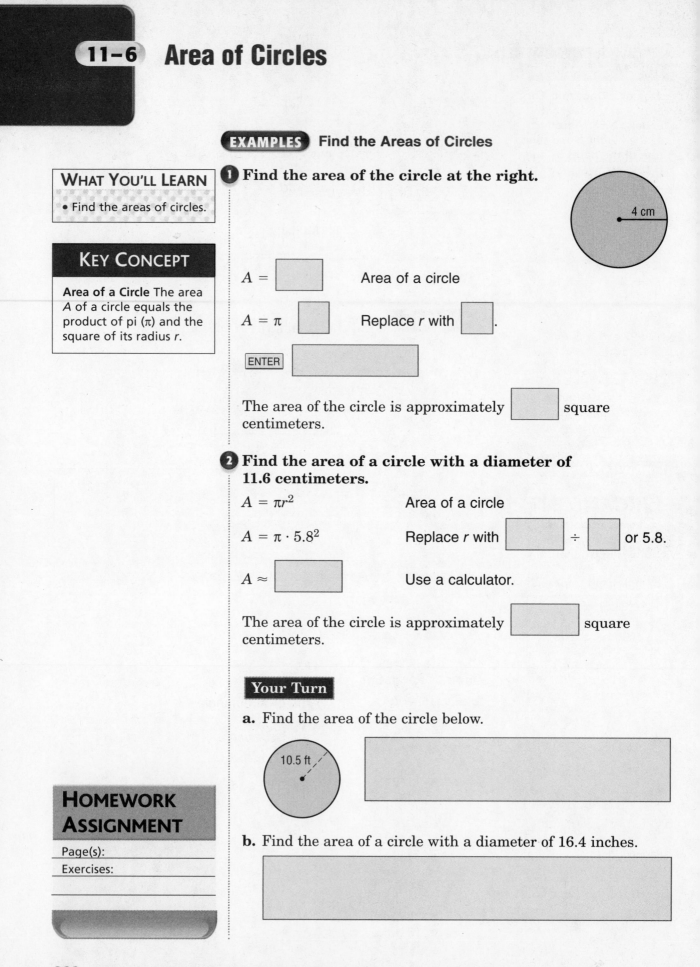

**EXAMPLES** Find the Areas of Circles

**1** Find the area of the circle at the right.

4 cm

$A = $ ▢     Area of a circle

$A = \pi$ ▢     Replace $r$ with ▢.

ENTER ▢

The area of the circle is approximately ▢ square centimeters.

**KEY CONCEPT**

**Area of a Circle** The area $A$ of a circle equals the product of pi ($\pi$) and the square of its radius $r$.

**2** Find the area of a circle with a diameter of 11.6 centimeters.

$A = \pi r^2$     Area of a circle

$A = \pi \cdot 5.8^2$     Replace $r$ with ▢ ÷ ▢ or 5.8.

$A \approx$ ▢     Use a calculator.

The area of the circle is approximately ▢ square centimeters.

**Your Turn**

**a.** Find the area of the circle below.

10.5 ft

**HOMEWORK ASSIGNMENT**

Page(s): _____

Exercises: _____

**b.** Find the area of a circle with a diameter of 16.4 inches.

# Area of Complex Figures

**BUILD YOUR VOCABULARY** (page 254)

A **complex figure** is made of [    ], rectangles,

[    ], and/or other [    ] figures.

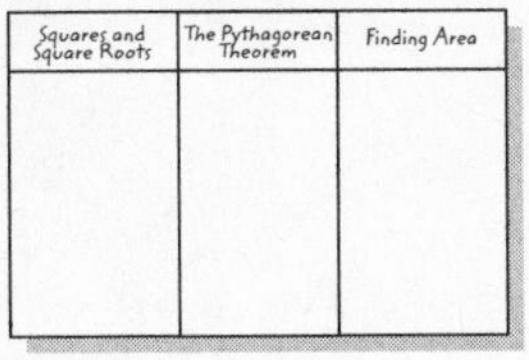

**EXAMPLE** Find the Area of an Irregular Room

1. **WINDOWS** The diagram at the right shows the dimensions of a window that is 3.4 feet by 7.2 feet. Find the area of the window. Round to the nearest tenth.

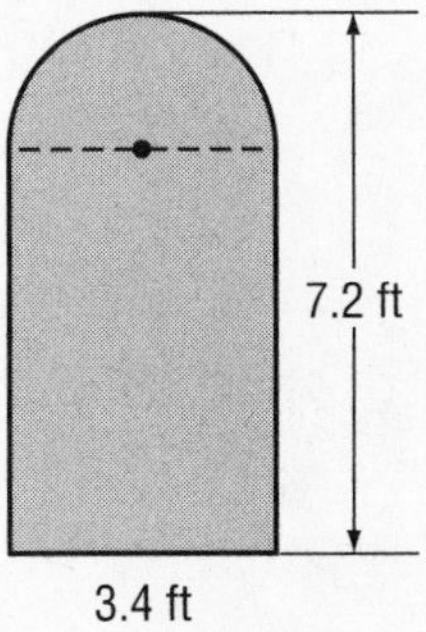

The figure can be separated into a semicircle and a rectangle.

7.2 ft

3.4 ft

Area of Semicircle

$A = $ [    ] $\pi r^2$     Area of a semicircle

$A = \frac{1}{2}\pi$ [    ]     Replace $r$ with [    ] ÷ [    ] or [    ].

$A \approx$ [    ]     Simplify.

Area of Rectangle

$A = \ell w$     Area of a rectangle

$A = $ [    ]     Replace $\ell$ with [    ] − [    ] or [    ]

and $w$ with [    ].

$A = $ [    ]     Multiply.

The area of the window is approximately [    ] + [    ]

or [    ] square feet.

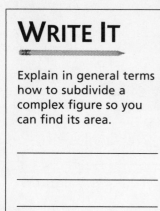

## WRITE IT

Explain in general terms how to subdivide a complex figure so you can find its area.

_____

_____

_____

_____

**Your Turn** The diagram below shows the dimensions of a new driveway. Find the area of the driveway. Round to the nearest tenth.

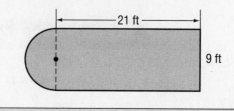

# Area Models and Probability

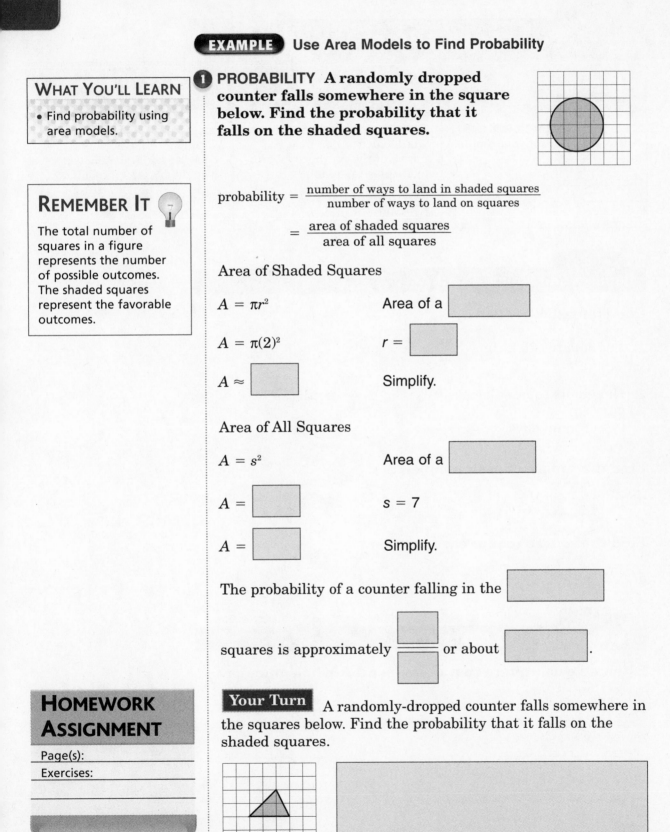

**EXAMPLE** Use Area Models to Find Probability

**WHAT YOU'LL LEARN**

• Find probability using area models.

**1** **PROBABILITY** A randomly dropped counter falls somewhere in the square below. Find the probability that it falls on the shaded squares.

**REMEMBER IT**

The total number of squares in a figure represents the number of possible outcomes. The shaded squares represent the favorable outcomes.

$$\text{probability} = \frac{\text{number of ways to land in shaded squares}}{\text{number of ways to land on squares}}$$

$$= \frac{\text{area of shaded squares}}{\text{area of all squares}}$$

Area of Shaded Squares

$A = \pi r^2$            Area of a 

$A = \pi(2)^2$          $r = $ 

$A \approx $            Simplify.

Area of All Squares

$A = s^2$              Area of a 

$A = $                $s = 7$

$A = $                Simplify.

The probability of a counter falling in the 

squares is approximately $=$ or about .

**Your Turn** A randomly-dropped counter falls somewhere in the squares below. Find the probability that it falls on the shaded squares.

**HOMEWORK ASSIGNMENT**

Page(s):

Exercises:

CHAPTER

# 11

# BRINGING IT ALL TOGETHER

## STUDY GUIDE

| **FOLDABLES**™ | VOCABULARY PUZZLEMAKER | **BUILD YOUR VOCABULARY** |
|---|---|---|
| Use your **Chapter 11 Foldable** to help you study for your chapter test. | To make a crossword puzzle, word search, or jumble puzzle of the vocabulary words in Chapter 11, go to:<br><br>www.glencoe.com/sec/math/ t_resources/free/index.php | You can use your completed **Vocabulary Builder** *(pages 254–255)* to help you solve the puzzle. |

### 11-1
### Squares and Square Roots

**Complete each sentence.**

1. The square of 3 means ☐ × ☐ .

2. Nine units squared means 9 ☐ with ☐ of
☐ unit each.

**Find the square of each number.**

3. 16 ☐                 4. 28 ☐

**Find the square root of each number.**

5. $\sqrt{121}$ ☐              6. $\sqrt{484}$ ☐

### 11-2
### Estimating Square Roots

**Estimate each square root to the nearest whole number.**

7. $\sqrt{95}$

8. $\sqrt{51}$

9. $\sqrt{150}$

10. $\sqrt{230}$

**11-3**

## The Pythagorean Theorem

**State whether each sentence is *true* or *false*. If false, replace the underlined word to make a true sentence.**

11. The Pythagorean Theorem states that $c^2 = a^2 + b^2$, where $\underline{a}$

    represents the length of the hypotenuse.

12. The <u>length</u> of a rectangle is the hypotenuse of two right

    triangles.

13. The <u>hypotenuse</u> is always the longest of the three sides of a

    right triangle.

14. A triangle with sides of length 8.1 centimeters, 10.8
    centimeters, and 13.5 centimeters <u>is</u> a right triangle.

**Find the missing measure of each right triangle. Round to the nearest tenth if necessary.**

15.

    3 in.

    8 in.

16.

    24 yd

    12 yd

**11-4**

## Area of Parallelograms

**State whether each sentence is *true* or *false*. If false, replace the underlined word to make a true sentence.**

17. To find the <u>base</u> of a parallelogram, draw a segment
    perpendicular to the base with endpoints on opposite sides of

    the parallelogram.

18. The area of a parallelogram is found by <u>multiplying</u> its base

    times the height.

19. What is the area of a parallelogram with a base of 15 feet and

    a height of 3.5 feet?

**11-5**

## Area of Triangles and Trapezoids

**Complete each sentence.**

**20.** The ⬚ of a triangle can be any of its ⬚.

**21.** To find the ⬚ of a triangle, find the distance from the ⬚ to the ⬚ vertex.

**Find the area.**

**22.** 

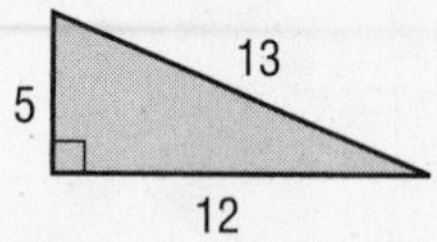

5, 13, 12 ⬚

**23.** 

9 in., 7 in., 13 in. ⬚

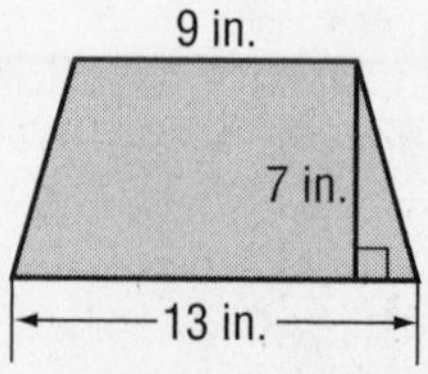

**11-6**

## Area of Circles

**Complete each sentence.**

**24.** To find the ⬚ of a circle when you are given the ⬚, divide the length of the diameter by ⬚, square that, and ⬚ the result by pi.

**25.** The units for the ⬚ of a circle will always be measured in ⬚ units.

**26.** Find the area of a circle with a diameter of 13.6 inches. Round to the nearest tenth. ⬚

**11-7**

## Area of Complex Figures

**Name the two dimensions of the following figures.**

**27.** rectangle ⬚

**28.** triangle ⬚

**Find the area of each figure. Round to the nearest tenth if necessary.**

29.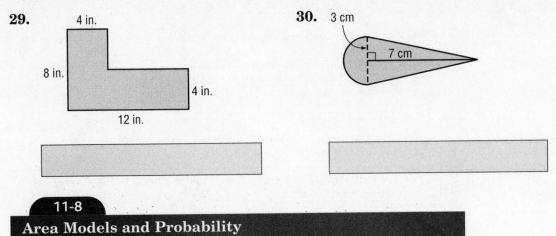

4 in.

8 in.

4 in.

12 in.

30. 3 cm

7 cm

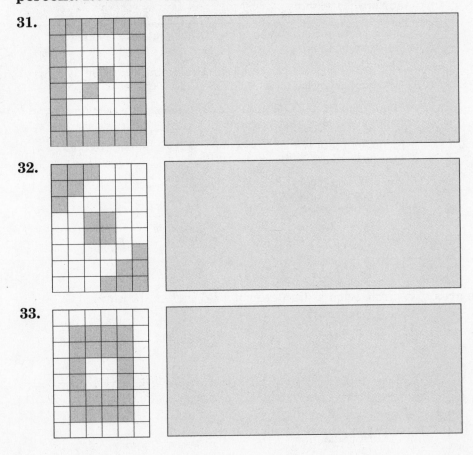

## 11-8
## Area Models and Probability

**A randomly-dropped counter falls in the squares. Find the probability that it falls in the shaded squares. Write as a percent. Round to the nearest tenth if necessary.**

31.

32.

33.

# ARE YOU READY FOR THE CHAPTER TEST?

Visit **msmath2.net** to access your textbook, more examples, self-check quizzes, and practice tests to help you study the concepts in Chapter 11.

**Check the one that applies. Suggestions to help you study are given with each item.**

☐ **I completed the review of all or most lessons without using my notes or asking for help.**

- You are probably ready for the Chapter Test.

- You may want to take the Chapter 11 Practice Test on page 507 of your textbook as a final check.

☐ **I used my Foldable or Study Notebook to complete the review of all or most lessons.**

- You should complete the Chapter 11 Study Guide and Review on pages 504–506 of your textbook.

- If you are unsure of any concepts or skills, refer back to the specific lesson(s).

- You may want to take the Chapter 11 Practice Test on page 507 of your textbook.

☐ **I asked for help from someone else to complete the review of all or most lessons.**

- You should review the examples and concepts in your Study Notebook and Chapter 11 Foldable.

- Then complete the Chapter 11 Study Guide and Review on pages 504–506 of your textbook.

- If you are unsure of any concepts or skills, refer back to the specific lesson(s).

- You may also want to take the Chapter 11 Practice Test on page 507 of your textbook.

Student Signature

Parent/Guardian Signature

Teacher Signature

# Geometry: Measuring Three-Dimensional Figures

**FOLDABLES™** Use the instructions below to make a Foldable to help you organize your notes as you study the chapter. You will see Foldable reminders in the margin of this Interactive Study Notebook to help you in taking notes.

**Begin with a piece of 11" by 17" paper.**

**STEP 1**  **Fold**
Fold the paper in fourths lengthwise.

**STEP 2**  **Open and Fold**
Fold a 2" tab along the short side. Then fold the rest in half.

**STEP 3**  **Label**
Draw lines along folds and label as shown.

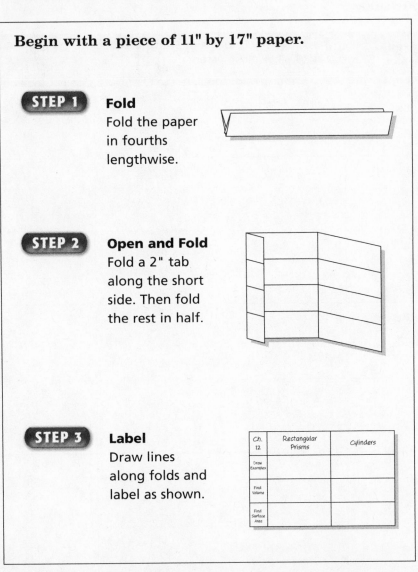

| Ch. 12 | Rectangular Prisms | Cylinders |
|---|---|---|
| Draw Examples | | |
| Find Volume | | |
| Find Surface Area | | |

**NOTE-TAKING TIP:** When taking notes about 3-dimensional figures, it is important to draw examples. It also helps to record any measurement formulas.

**BUILD YOUR VOCABULARY**

This is an alphabetical list of new vocabulary terms you will learn in Chapter 12. As you complete the study notes for the chapter, you will see Build Your Vocabulary reminders to complete each term's definition or description on these pages. Remember to add the textbook page number in the second column for reference when you study.

| Vocabulary Term | Found on Page | Definition | Description or Example |
|---|---|---|---|
| cylinder [SILL-in-der] | | | |
| precision [pri-SIH-zhuhn] | | | |
| precision unit | | | |
| rectangular prism | | | |
| significant digits | | | |
| solid | | | |
| surface area | | | |
| volume | | | |

# Drawing Three-Dimensional Figures

(page 276)

## WHAT YOU'LL LEARN

• Draw a three-dimensional figure given the top, side, and front views.

**BUILD YOUR VOCABULARY** (page 276)

A **solid** is a [_____] figure because it has

length, width, and [_____].

**FOLDABLES**

## ORGANIZE IT

Record notes about drawing three-dimensional figures in the appropriate sections of your Foldable table. Sketch examples of rectangular prisms and cylinders.

| Ch. 12 | Rectangular Prisms | Cylinders |
|---|---|---|
| Draw Examples | | |
| Find Volume | | |
| Find Surface Area | | |

**EXAMPLE** Draw Different Views of a Solid

**1 Draw a top, a side, and a front view of the figure below.**

The top and front views are [_____]. The side

view is a [_____].

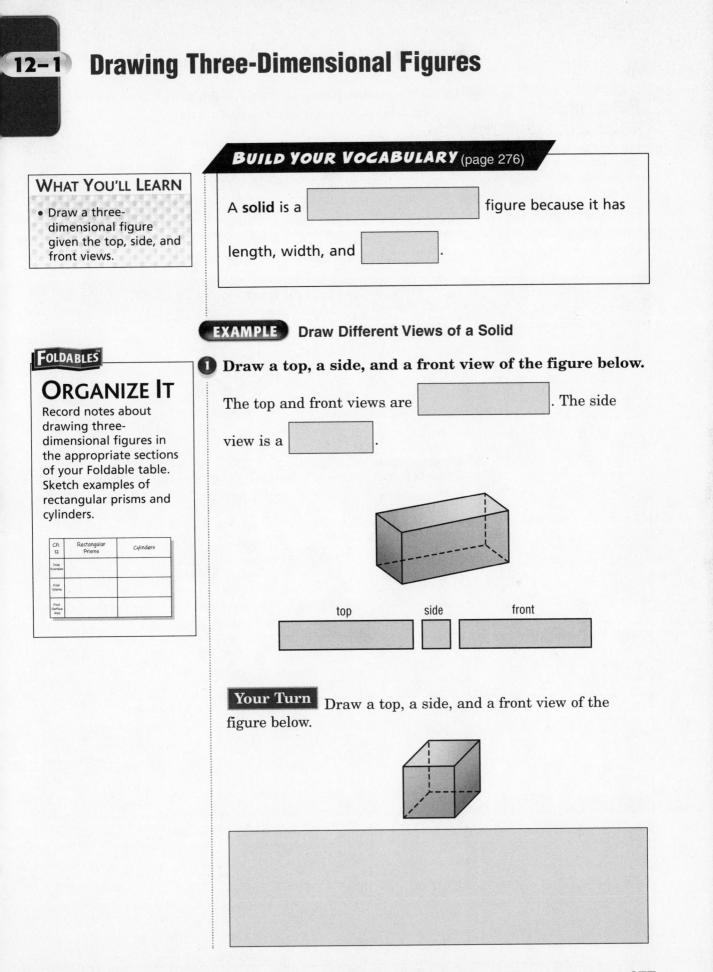

top    side    front

**Your Turn** Draw a top, a side, and a front view of the figure below.

**REMEMBER IT**

There is more than one way to draw the different views of a three-dimensional figure.

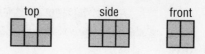

**EXAMPLE** Draw a Three-Dimensional Figure

**2** Draw the solid using the top, side, and front views shown below. Use isometric dot paper.

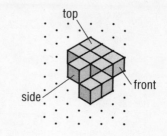

top      side      front

**Step 1**    Use the top view to draw the base of the figure.

**Step 2**    Add edges to make the base a solid figure.

**Step 3**    Use the side and front views to complete the figure.

**Your Turn**  Draw a solid using the top, side, and front views shown below. Use isometric dot paper.

top      side      front

**HOMEWORK ASSIGNMENT**

Page(s):

Exercises:

# Volume of Rectangular Prisms

## WHAT YOU'LL LEARN

- Find the volumes of rectangular prisms.

A **volume** of a solid is the measure of [          ] occupied by it.

A **rectangular prism** is a solid figure that has two

[          ] and congruent sides, or bases, that are

[          ].

## KEY CONCEPT

**Volume of a Rectangular Prism** The volume $V$ of a rectangular prism is the area of the base $B$ times the height $h$. It is also the product of the length $\ell$, the width $w$, and the height $h$.

**EXAMPLE**  Find the Volumes of Prisms

1. Find the volume of the rectangular prism.

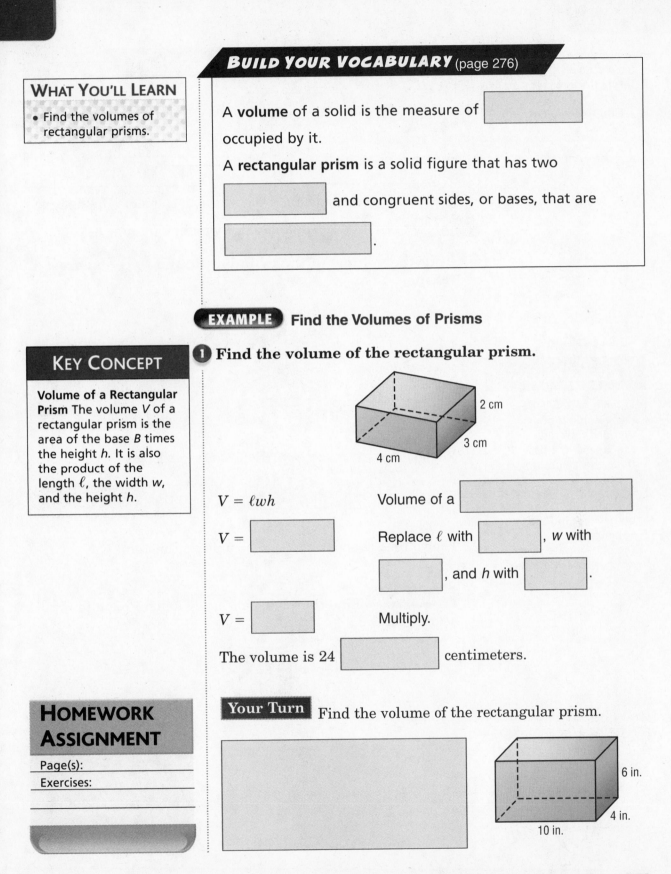

2 cm
3 cm
4 cm

$V = \ell wh$      Volume of a [          ]

$V =$ [          ]      Replace $\ell$ with [          ], $w$ with

[          ], and $h$ with [          ].

$V =$ [          ]      Multiply.

The volume is 24 [          ] centimeters.

## HOMEWORK ASSIGNMENT

Page(s):

Exercises:

**Your Turn**  Find the volume of the rectangular prism.

6 in.
4 in.
10 in.

# Volume of Cylinders

## WHAT YOU'LL LEARN

- Find the volumes of cylinders.

(page 276)

A **cylinder** is a [        ] figure that has two

[        ], parallel [        ] as its bases.

---

**EXAMPLE** Find the Volume of a Cylinder

**1** Find the volume of the cylinder. Round to the nearest tenth.

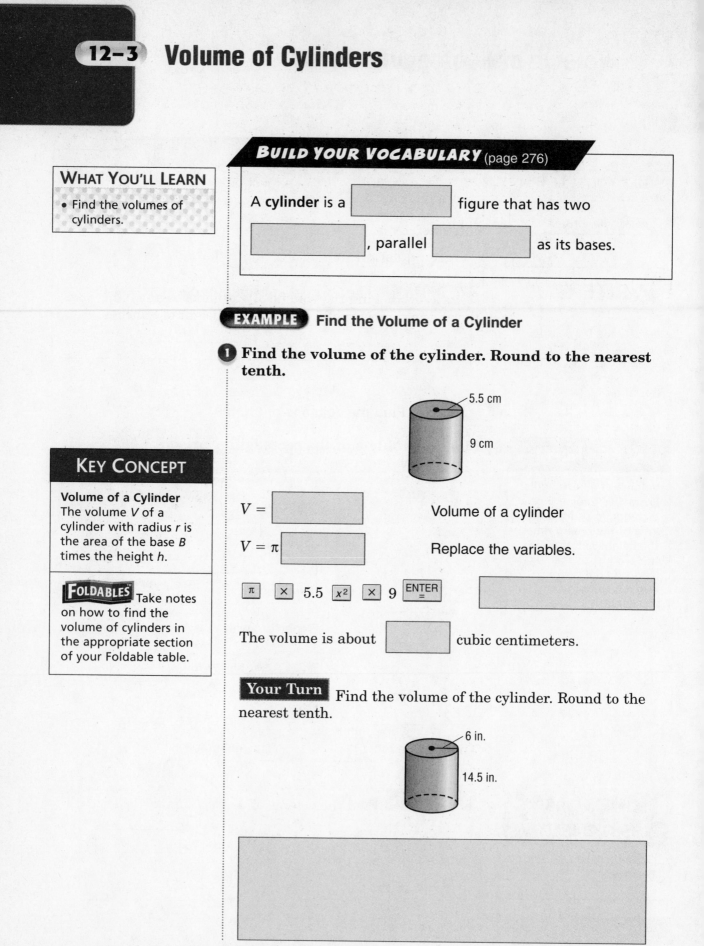

5.5 cm

9 cm

$V =$ [        ]        Volume of a cylinder

$V = \pi$ [        ]        Replace the variables.

π  ×  5.5  $x^2$  ×  9  ENTER =        [        ]

The volume is about [        ] cubic centimeters.

## KEY CONCEPT

**Volume of a Cylinder**
The volume $V$ of a cylinder with radius $r$ is the area of the base $B$ times the height $h$.

**FOLDABLES** Take notes on how to find the volume of cylinders in the appropriate section of your Foldable table.

**Your Turn** Find the volume of the cylinder. Round to the nearest tenth.

6 in.

14.5 in.

---

**EXAMPLE** Find the Volume of a Real-Life Object

2 **Find the volume of a cylinder-shaped coffee can that has a diameter of 3 inches and a height of 6 inches.**

$V = \pi r^2 h$          Volume of a cylinder

$V = \pi \boxed{\phantom{xxxxxxxxx}}$      Replace the variables.

$V = \boxed{\phantom{xxxx}}$         Simplify.

**Your Turn** Find the volume of a cylinder-shaped juice can that has a diameter of 5 inches and a height of 8 inches.

**Surface Area of Rectangular Prisms**

## WHAT YOU'LL LEARN

• Find the surface areas of rectangular prisms.

**BUILD YOUR VOCABULARY** (page 276)

The ⬚ of the areas of all of the ⬚, or faces, of a ⬚ figure is the surface area.

**EXAMPLE** Use a Net to Find Surface Area

## KEY CONCEPT

**Surface Area of Rectangular Prisms** The surface area $S$ of a rectangular prism with length $\ell$, width $w$, and height $h$ is the sum of the areas of the faces.

① **Find the surface area of the rectangular prism.**

You can use a net of the rectangular prism to find its surface area. There are three pairs of congruent faces.

• top and bottom

• front and back

• two sides

| Faces | Area | |
|---|---|---|
| top and bottom | $2(6 \cdot 2) =$ | ⬚ |
| front and back | $2(6 \cdot 3) =$ | ⬚ |
| two sides | $2(2 \cdot 3) =$ | ⬚ |

The surface area is ⬚ + ⬚ + ⬚ or ⬚ square centimeters.

**Your Turn** Find the surface area of the rectangular prism.

12-4

**EXAMPLE** Use a Formula to Find Surface Area

**2** Find the surface area of the rectangular prism.

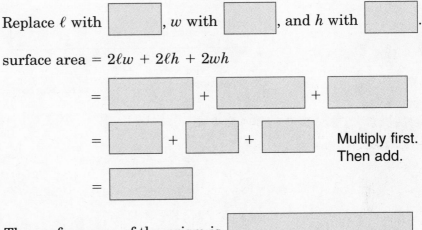

Replace $\ell$ with ⬚, $w$ with ⬚, and $h$ with ⬚.

$= $ ⬚ $+$ ⬚ $+$ ⬚   Multiply first. Then add.

The surface area of the prism is ⬚.

**Your Turn** Find the surface area of the rectangular prism.

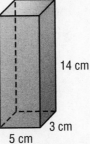

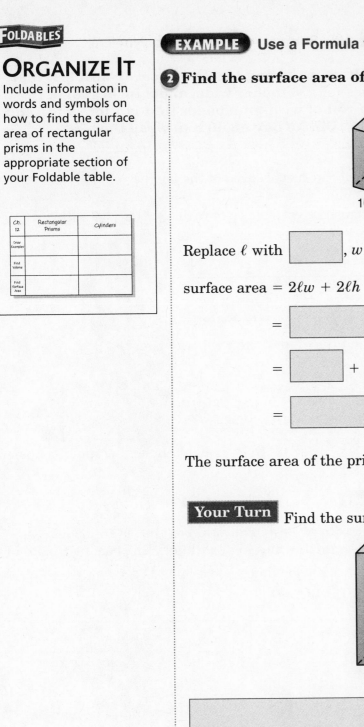

**FOLDABLES**

**ORGANIZE IT**

Include information in words and symbols on how to find the surface area of rectangular prisms in the appropriate section of your Foldable table.

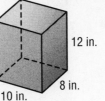

**EXAMPLE**  Use Surface Area to Solve a Problem

**3** **BOXES** Drew is putting together a cardboard box that is 9 inches long, 6 inches wide, and 8 inches high. He bought a roll of wrapping paper that is 1 foot wide and 3 feet long. Did he buy enough to wrap the box? Explain.

**Step 1** Find the surface area of the box.

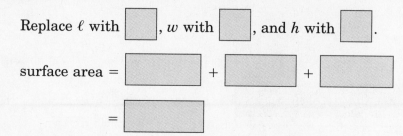

Replace $\ell$ with [ ], $w$ with [ ], and $h$ with [ ].

surface area = [ ] + [ ] + [ ]

= [ ]

**Step 2** Find the area of the wrapping paper.

[ 1 ft ]        [ 3 ft ]

area = 12 in. · 36 in. or 432 in²

Since 432 [ ] 348, Drew bought enough wrapping paper.

**Your Turn**

Angela needs to cover a cardboard box that is 15 inches long, 5 inches wide, and 4 inches high with felt. She bought a piece of felt that is 1 foot wide and $2\frac{1}{2}$ feet long. Did she buy enough felt to cover the box? Explain.

**HOMEWORK
ASSIGNMENT**

Page(s):

Exercises:

# Surface Area of Cylinders

### WHAT YOU'LL LEARN

- Find the surface areas of cylinders.

### KEY CONCEPT

**Surface Area of a Cylinder** The surface area *S* of a cylinder with height *h* and radius *r* is the sum of the areas of circular bases and the area of the curved surface.

**EXAMPLE** Find Surface Area of a Cylinder

❶ **Find the surface area of the cylinder. Round to the nearest tenth.**

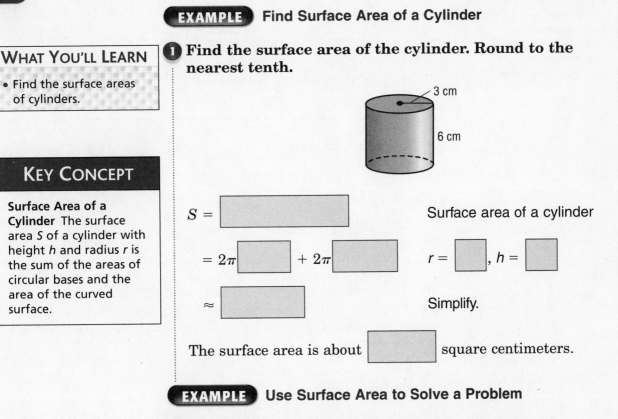

3 cm

6 cm

$S =$ ⬜                    Surface area of a cylinder

$= 2\pi$ ⬜ $+ 2\pi$ ⬜           $r =$ ⬜ , $h =$ ⬜

$\approx$ ⬜                    Simplify.

The surface area is about ⬜ square centimeters.

**EXAMPLE** Use Surface Area to Solve a Problem

❷ **GIFT WRAP A poster is contained in a cardboard cylinder that is 10 inches high. The cylinder's base has a diameter of 8 inches. How much paper is needed to wrap the cardboard cylinder if the ends are to be left uncovered?**

Since only the curved side of the cylinder is to be covered, you do not need to include the areas of the top and bottom of the cylinder.

$S =$ ⬜                    Curved surface of a cylinder

$=$ ⬜                    $r = 4, h = 10$

$\approx$ ⬜                    Simplify.

About 251.3 ⬜ of paper is needed.

## ORGANIZE IT

Include information in words and symbols about how to find the surface area of a cylinder in the appropriate section of your Foldable table.

| Ch. 12 | Rectangular Prisms | Cylinders |
|---|---|---|
| Draw Examples | | |
| Find Volume | | |
| Find Surface Area | | |

**Your Turn**

**a.** Find the surface area of the cylinder. Round to the nearest tenth.

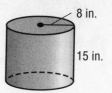

8 in.

15 in.

**b.** A can of fruit juice is in the shape of a cylinder with a diameter of 6 inches and a height of 12 inches. How much paper is needed to create the label if the ends are to be left uncovered?

## HOMEWORK ASSIGNMENT

Page(s):

Exercises:

# Precision and Measurement

## WHAT YOU'LL LEARN

- Determine and apply significant digits in a real-life context.

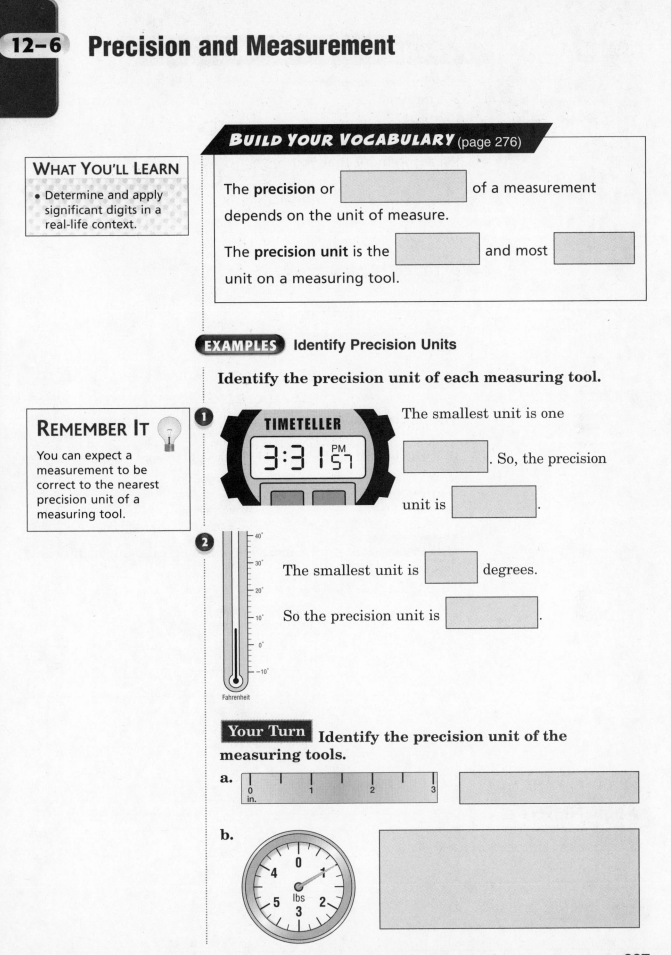

### BUILD YOUR VOCABULARY (page 276)

The **precision** or _____ of a measurement depends on the unit of measure.

The **precision unit** is the _____ and most _____ unit on a measuring tool.

**EXAMPLES** Identify Precision Units

**Identify the precision unit of each measuring tool.**

## REMEMBER IT

You can expect a measurement to be correct to the nearest precision unit of a measuring tool.

**1**

TIMETELLER

3:31 57 PM

The smallest unit is one _____. So, the precision unit is _____.

**2**

The smallest unit is _____ degrees.

So the precision unit is _____.

Fahrenheit

**Your Turn** Identify the precision unit of the measuring tools.

**a.**

0   1   2   3
in.

**b.**

lbs

**BUILD YOUR VOCABULARY** (page 276)

A more [                ] method of measurement is that of **significant digits**, which include all of the digits of a measurement that you [                ] for sure, plus one [                ] digit.

**EXAMPLE** Apply Significant Digits

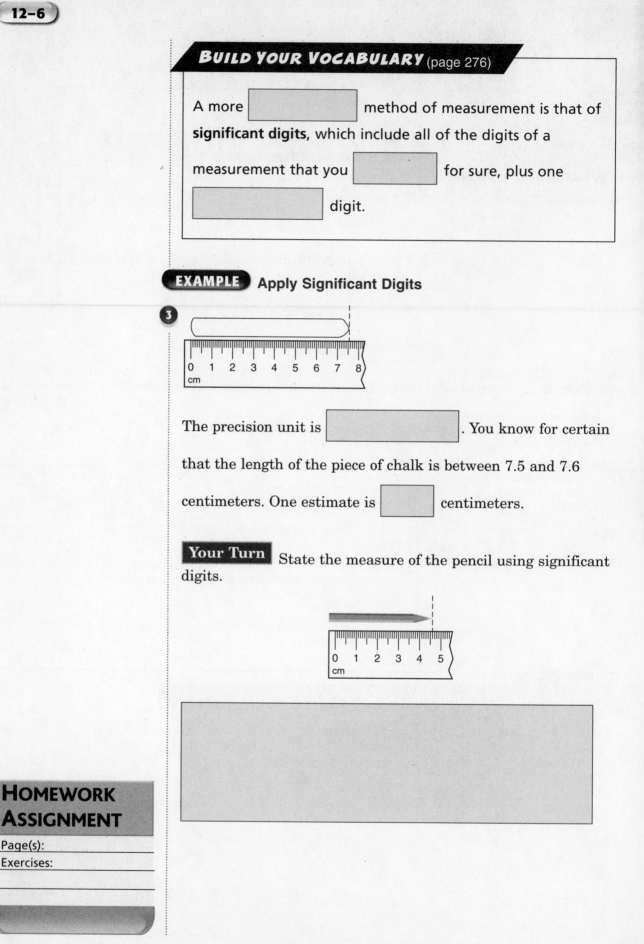

**3**

The precision unit is [                ]. You know for certain that the length of the piece of chalk is between 7.5 and 7.6 centimeters. One estimate is [          ] centimeters.

**Your Turn** State the measure of the pencil using significant digits.

**HOMEWORK ASSIGNMENT**

Page(s): _____

Exercises: _____

# BRINGING IT ALL TOGETHER

## STUDY GUIDE

| FOLDABLES™ | VOCABULARY PUZZLEMAKER | BUILD YOUR VOCABULARY |
|---|---|---|
| Use your **Chapter 12 Foldable** to help you study for your chapter test. | To make a crossword puzzle, word search, or jumble puzzle of the vocabulary words in Chapter 12, go to:<br><br>www.glencoe.com/sec/math/ t_resources/free/index.php | You can use your completed **Vocabulary Builder** *(page 276)* to help you solve the puzzle. |

### 12-1
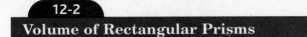

## Drawing Three-Dimensional Figures

**Complete each sentence.**

1. A two-dimensional figure has two dimensions; [          ] and

   [          ].

2. A three-dimensional figure has three dimensions; [          ],

   [          ] and [          ].

**Underline the word that makes the sentence true.**

3. A (rectangle, cube) is a three dimensional figure.

### 12-2
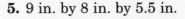

## Volume of Rectangular Prisms

**Find the volume of rectangular prisms with these dimensions. Round to the nearest tenth if necessary.**

4. 4 ft by 12 ft by 7 ft

   [          ]

5. 9 in. by 8 in. by 5.5 in.

   [          ]

6. 2.5 in. by 6 in. by 5 in.

7. 3.8 cm by 2.4 cm by 2 cm

**12-3**

## Volume of Cylinders

Write C if the phrase is true of a cylinder, P if it is true of a prism, and CP if the phrase is true of both.

8. [    ] has bases that are parallel and congruent.

9. [    ] has sides and bases that are polygons.

10. [    ] has bases that are circular.

11. [    ] is a solid.

12. [    ] has volume.

13. [    ] is three-dimensional.

**12-4**

## Surface Area of Rectangular Prisms

Find the surface area of each rectangular prism. Round to the nearest tenth if necessary.

14.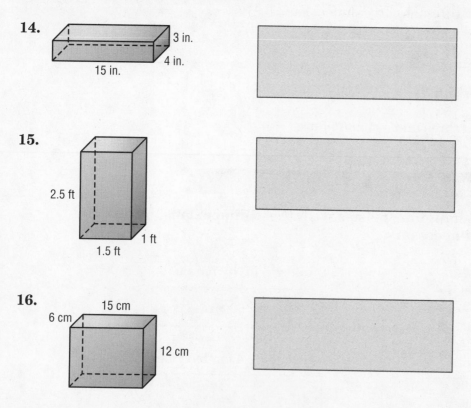

15.

16.

**12-5**

## Surface Area of Cylinders

**Write the formula to find each of the following.**

**17.** the area of a circle

**18.** the circumference of a circle

**19.** the area of a rectangle

**Find the surface area of the cylinder. Round to the nearest tenth if necessary.**

**20.**

3 in.

10 in.

**12-6**

## Precision and Measurement

**State whether each sentence is true or false. If false, replace the underlined word to make a true sentence.**

**21.** It might be necessary to use significant digits when an exact measure is <u>known</u>.

**22.** When you use significant degrees, <u>two</u> digit(s) is/are estimated.

**23.** A ruler that has markings for centimeters is <u>more</u> precise than a ruler with markings for millimeters.

**24.** If you use a stopwatch with a precision unit of measure of $\frac{1}{100}$, the measurement will still be an <u>approximation</u> since all measurements could be more precise.

# ARE YOU READY FOR THE CHAPTER TEST?

**Check the one that applies. Suggestions to help you study are given with each item.**

☐ **I completed the review of all or most lessons without using my notes or asking for help.**

- You are probably ready for the Chapter Test.

- You may want to take the Chapter 12 Practice Test on page 549 of your textbook as a final check.

☐ **I used my Foldable or Study Notebook to complete the review of all or most lessons.**

- You should complete the Chapter 12 Study Guide and Review on pages 546–548 of your textbook.

- If you are unsure of any concepts or skills, refer back to the specific lesson(s).

- You may want to take the Chapter 12 Practice Test on page 549 of your textbook.

☐ **I asked for help from someone else to complete the review of all or most lessons.**

- You should review the examples and concepts in your Study Notebook and Chapter 12 Foldable.

- Then complete the Chapter 12 Study Guide and Review on pages 546–548 of your textbook.

- If you are unsure of any concepts or skills, refer back to the specific lesson(s).

- You may also want to take the Chapter 12 Practice Test on page 549 of your textbook.

Student Signature

Parent/Guardian Signature

Teacher Signature